27.95

5/13

YA

333.92

D0602086

This book was brought
to you by the

Naumes Family Foundation

YUBA COUNTY LIBRARY
MARYSVILLE

The Future of Renewable Energy

What Is the Future of Wind Power?

by Andrea C. Nakaya

ReferencePoint Press®

San Diego, CA

About the Author

Andrea C. Nakaya, a native of New Zealand, holds a BA in English and an MA in communications from San Diego State University. She currently lives in Encinitas, California, with her husband and their two children, Natalie and Shane.

© 2013 ReferencePoint Press, Inc.
Printed in the United States

For more information, contact:
ReferencePoint Press, Inc.
PO Box 27779
San Diego, CA 92198
www.ReferencePointPress.com

ALL RIGHTS RESERVED.
No part of this work covered by the copyright hereon may be reproduced or used in any form or by any means—graphic, electronic, or mechanical, including photocopying, recording, taping, web distribution, or information storage retrieval systems—without the written permission of the publisher.

Picture Credits:
Cover: Thinkstock.com
Maury Aaseng: 11, 17, 23, 30, 37, 42, 48, 55, 62
Thinkstock.com: 9

LIBRARY OF CONGRESS CATALOGING-IN-PUBLICATION DATA

Nakaya, Andrea C., 1976–
  What is the future of wind power? / by Andrea C. Nakaya.
    p. cm. — (Future of renewable energy series)
  Includes bibliographical references and index.
  ISBN-13: 978-1-60152-280-1 (hardback)
  ISBN-10: 1-60152-280-0 (hardback)
  1. Wind power. 2. Energy development. I. Title.
  TJ820.N35 2013
  333.9'2—dc23
                                                        2012000281

Contents

Foreword

What are the long-term prospects for renewable energy?

In his 2011 State of the Union address, President Barack Obama set an ambitious goal for the United States: to generate 80 percent of its electricity from clean energy sources, including renewables such as wind, solar, biomass, and hydropower, by 2035. The president reaffirmed this goal in the March 2011 White House report *Blueprint for a Secure Energy Future*. The report emphasizes the president's view that continued advances in renewable energy are an essential piece of America's energy future. "Beyond our efforts to reduce our dependence on oil," the report states, "we must focus on expanding cleaner sources of electricity, including renewables like wind and solar, as well as clean coal, natural gas, and nuclear power—keeping America on the cutting edge of clean energy technology so that we can build a 21st century clean energy economy and win the future."

Obama's vision of America's energy future is not shared by all. Benjamin Zycher, a visiting scholar at the American Enterprise Institute, a conservative think tank, contends that policies aimed at shifting from conventional to renewable energy sources demonstrate a "disconnect between the rhetoric and the reality." In *Renewable Electricity Generation: Economic Analysis and Outlook* Zycher writes that renewables have inherent limitations that can be overcome only at a very high cost. He states: "Renewable electricity has only a small share of the market, and ongoing developments in the market for competitive fuels . . . make it likely that renewable electricity will continue to face severe constraints in terms of competitiveness for many years to come."

Is Obama's goal of 80 percent clean electricity by 2035 realistic? Expert opinions can be found on both sides of this question and on all of the other issues relating to the debate about what lies ahead for renewable energy. Driven by this reality, the Future of Renewable Energy series critically examines the long-term prospects for renewable energy by delving into the topics and opinions that dominate and inform renewable energy policy and debate. The series covers renewables such as solar, wind, biofuels, hydrogen, and hydropower and explores the issues of cost and affordability, impact on the environment, viability as a replacement for fossil fuels, and what role—if any—government should play in renewable energy development. Pointed questions (such as "Can Solar Power Ever Replace Fossil Fuels?" or "Should Government Play a Role in Developing Biofuels?") frame the discussion and support inquiry-based learning. The pro/con format of the series encourages critical analysis of the topics and opinions that shape the debate. Discussion in each book is supported by current and relevant facts and illustrations, quotes from experts, and real-world examples and anecdotes. Additionally, all titles include a list of useful facts, organizations to contact for further information, and other helpful sources for further reading and research.

Visions of the Future: Wind Power

El Hierro is a tiny Spanish island with a population of about 10,000 people. It has no coal, natural gas, or oil of its own, and until recently has depended on importing these fossil fuels to supply its electricity. The island plans to change this with wind power. In 2012 El Hierro was working to become the first inhabited landmass in the world that is completely energy self-sufficient. Electricity for the island was to be generated by five large wind turbines. Surplus wind power was to be used to pump water uphill to a large reservoir. When the wind stops blowing or demand is too high, water was to be released from the reservoir to create hydroelectric power. So, through a combination of wind and hydroelectric power, El Hierro was to provide a continuous supply of electricity to its residents.

An Increasingly Important Source of Electricity

Most of the world's electricity is currently generated by burning fossil fuels. However, many people believe wind power is a cleaner, more sustainable, and ultimately more affordable alternative. El Hierro was hoping to prove that, with innovation, wind power can be used to replace fossil fuels. Wind power advocates hope that this project will lead the way for many more like it and that in the future, wind power will generate a large amount of the world's electricity. Says Tomás Padrón, leader of the island council, "When we started telling officials in Madrid back in the 1980s about our project, they thought it was a utopia and they turned their backs on us. Now it's a world-class model for all."[1] Not only will wind power save El Hierro an estimated $4 million a year that was previously spent on oil imports, according to a 2011 *New York Times* report, but the island will actually generate millions of dollars in revenue from selling its electricity to consumers.

Overall, wind power still makes up a very small part of global electricity generation. However, many countries are setting goals to generate significant percentages of their electricity from wind in the next few decades. According to the World Wind Energy Association, at the end of 2010 all the wind turbines installed worldwide were capable of generating only 2.5 percent of global electricity needs. However, this is more than double that of only a few years before, and many countries have ambitious plans to substantially increase their wind power production in the near future. The European Wind Energy Association predicts that by 2020, Europe will meet 15.7 percent of its electrical demand with wind power. The association hopes that by 2050, that amount will increase to 50 percent. According to a 2011 report, *China Wind Energy Development Roadmap 2050*, China could produce 17 percent of its electricity through wind power by 2050. And in the United States, the US Department of Energy (DOE) estimates that it is technologically feasible for the country to provide 20 percent of its electrical demand with wind energy by 2030.

An Abundant Source of Power

Wind power is based on a simple concept, where spinning turbines capture the energy of the wind and convert it into a usable form of power. Hundreds of years ago, people harnessed this power through windmills and used it to pump water and grind grain. Today turbines are used to convert wind energy into electricity. As a turbine spins, it powers a generator. The generator sends electricity into a utility grid, where it is distributed to customers just like the electricity from conventional fossil fuel power plants. Wind turbines range in size from small ones that provide enough power for just one home or business to huge structures with blades longer than the length of a football field. The latter, standing 20 stories high, can produce enough electricity to power more than 1,000 homes. Often many turbines are grouped together in a wind farm. One of the world's largest onshore wind farms is located in West Texas. Comprised of 627 wind turbines and covering more than 100,000 acres, the Roscoe Wind Farm provides enough power for about 250,000 homes.

Widespread use of wind turbines for electricity began about 30 years ago and since then there have been significant technological improvements. Early turbines were plagued with problems. Explains the *New York Times*, "Some came apart in bad storms, some did not work well, even in good weather, and still others found insects piling up on the blades, slowing power production."[2] Wind turbines have changed a lot since then, becoming more reliable, larger, and more efficient. According to Finn Strom Madsen, president of technology research and development for the Danish turbine manufacturer Vestas, one of the models made by his company can produce 300 times as much power as a turbine from 15 years ago.

Wind power is an attractive option for electricity production because wind is an abundant source of power that does not have many of the negative social and environmental impacts that fossil fuels do. The US Department of the Interior Bureau of Land Management explains some of the advantages of wind: "Wind energy avoids the external or societal costs associated with conventional resources, namely, the trade deficit from importing foreign oil and other fuels, the health and environmental costs of pollution, and the cost of depleted resources. Wind energy is a domestic, reliable resource that provides more jobs per dollar invested than any other energy technology."[3] Because of these advantages, advocates argue that wind should be used to replace as much fossil fuel power as possible.

Wind Power Around the World

Many countries around the world already generate significant amounts of electricity from the wind. According to a 2011 report prepared by the Lawrence Berkeley National Laboratory for the DOE, at the end of 2010, wind power was capable of supplying about 26 percent of Denmark's electricity, 17 percent of Portugal's, 15 percent of Spain's, 14 percent of Ireland's, and 9 percent of Germany's. The United States has the capacity to generate about 2.9 percent of its electricity from the wind.

Wind power currently supplies only a small portion of the world's electricity needs but in the coming decades many countries plan to increase their use of wind power, which is considered a clean and sustainable source of energy.

Although the United States is not a world leader in the percentage of its total electricity generated from the wind, it is one of the leaders in total installed wind power capacity—which is the total amount of wind power that its turbines can produce. The World Wind Energy Association reports that by the end of 2010, the countries with the highest installed wind power capacity were China, with 44,733 megawatts (MW); the United States, with 40,180 MW; Germany, with 27,215 MW; and Spain, with 20,676 MW.

"Wind energy avoids the external or societal costs associated with conventional resources."[3]

— US Department of the Interior Bureau of Land Management, a government agency that administers America's public lands.

In the last decade the United States has experienced unprecedented growth in wind power. According to the Department of Energy, it broke its own installation records in 2007, 2008, and 2009. In 2008 it surpassed Germany to become the world leader in installed wind power capacity. However, in 2010 growth dropped to about half the previous year due to economic recession, and the United States fell to second place behind China in installed wind power capacity. Texas is the US state with the greatest capacity, followed by Iowa, then California. Texas generates about 7.8 percent of its electricity with wind power, according to the Department of Energy, while in Iowa almost 20 percent of power consumption is supplied by the wind.

Obstacles to Widespread Development

To achieve even greater penetration in the electricity market, wind power still needs to overcome some significant obstacles. The biggest problem with wind is that, unlike fossil fuels, it does not create power in a steady and continuous supply. This is because the wind blows intermittently and at changing speeds. Cities using increasing amounts of wind power need to solve the problem of how to integrate this fluctuating power source into the power grid. They also need to find ways to provide continuous power to consumers, even when the wind is not blowing.

Wind Power Potential

The United States has many windy areas that could supply significant amounts of electricity if wind power were fully developed there. Some regions are better than others, however. Some of the windiest spots in the country are located in Montana, North Dakota, and Wyoming. These are categorized as class 5 and above, which means they experience some of the highest wind speeds in the United States, however these very windy areas are located relatively far from major population centers. On the other hand, class 3 and 4 wind speeds are more common and more evenly distributed around the country. Generating power economically relies on wind speeds in class 3 and above, which makes these areas desirable locations for the development of wind power.

Source: Union of Concerned Scientists, "How Wind Energy Works," December 15, 2009. www.ucsusa.org.

Wind Power Class	Resource Potential
3	Fair
4	Good
5	Excellent
6	Outstanding
7	Superb

Solving problems such as this will require significant financial investment. In addition, expanding wind power will necessitate costly new transmission lines to transport that power to consumers. There is disagreement over how much money should be invested in overcoming wind power's current limitations, and how much—if any—should be the responsibility of government.

The government currently provides significant support for the wind power industry through tax credits, grants, and mandates requiring the use of wind power and other renewable energy. However, such support is controversial. While wind advocates maintain that the industry needs this support to help it compete in the electricity market, critics contend that government involvement in wind power is actually impeding innovation.

An Important Resource for the Future

Despite these obstacles and areas of disagreement, many people believe that wind power will play a major role in meeting the world's future energy needs. As worldwide electricity demand continues to increase, it is widely recognized that society needs to pursue alternatives to fossil fuel electricity. Of all the existing alternatives, wind power has proved capable of producing large amounts of electricity with minimal harms to society and the environment, and it is one of the fastest-growing sources of energy around the world. Wind power, expert Peter Musgrove says, "Is now recognized as the most promising and cost-effective of the high-resource renewable energy options."[4]

Chapter One

Is Wind Power Affordable?

Wind Power Is Affordable

The price of wind power is low enough to be competitive with other energy sources such as coal and natural gas. In addition, wind will become increasingly competitive in the future as technology improves and as fossil fuels become more scarce and expensive. Wind has some other important advantages over fossil fuels that make it even more affordable. It does not pollute the environment, so it does not incur the large health-care and environmental damage costs that fossil fuels do. It also helps the US economy by creating local jobs. And because wind is a local source of energy and not imported, it keeps money spent on energy within the US economy instead of being spent overseas.

The Debate

Wind Power Is Too Costly

Compared to fossil fuels, wind power is simply too costly as a source of electricity. Although coal and natural gas do cause some environmental harms, these harms have been greatly exaggerated, and fossil fuels are still the most affordable way to provide electricity to consumers. Widespread use of wind power will necessitate the construction of hundreds of miles of new transmission lines and backup power plants to provide electricity when the wind is not blowing. These expenses make wind power even more costly for society.

Wind Power Is Affordable

"Wind is cost competitive with all other sources of new electricity."

—American Wind Energy Association, a trade association that represents the wind power industry.

American Wind Energy Association, "Clean. Affordable. Homegrown. American Wind Power," 2010. www.awea.org.

The cost of wind power is competitive with other sources of electricity and will become increasingly competitive in the future. Price estimates comparing wind and other forms of power generation vary significantly by region and by method of analysis. However, in sites with good wind resources, wind power can cost less than coal, natural gas, or other forms of energy production. In other areas the average price is low enough to be affordable and will become more and more competitive over time. In a 2011 report, the Center for Climate and Energy Solutions compares estimated costs of wind, coal, and natural gas power and finds that wind is cost-competitive with coal and natural gas in some regions of the United States. According to the report, the levelized cost of generating electricity with a new wind power project is 6–11 cents per kilowatt-hour. Coal-fueled electricity is 6.4–9.5 cents, and natural gas is 6.9–9.6 cents. Levelized cost is calculated by taking the estimated costs of building and operating the plant power over its lifetime and calculating the cost per unit of electricity produced. The Union of Concerned Scientists states, "Wind power is steadily becoming one of the most cost-effective choices for electricity in the United States."[5]

Some countries are proving this affordability by successfully generating significant—and increasing—percentages of their electricity from wind. For example, Denmark currently gets about 20 percent of its energy from wind and hopes to get 50 percent by 2020. The Danish Wind Industry Association states that the production cost per kilowatt-hour

14

for wind energy has dropped more than 80 percent in the past 20 years. It predicts that this trend will continue, with wind power becoming fully competitive with other sources of electricity within 7 to 10 years. Portugal gets about 18 percent of its energy from wind power. As a result of Portugal's use of wind and other renewable energy sources, the country hopes to close at least 2 conventional power plants by 2014, and reduce the operation of others, according to a 2010 *New York Times* report.

Continually Decreasing Prices

Compared to traditional sources of energy, wind power is a relatively new technology, which means that it is still being developed and is likely to become more efficient and less costly over time. For example, there is great potential for improvement of wind turbine efficiency. Every turbine is rated according to the amount of energy it is capable of producing under ideal conditions. (Wind turbines never produce at their maximum capability because wind speed is never constant.) Currently, turbines produce less than half of their maximum capacity—and often significantly less. Improved turbine technology will allow turbines to capture a greater percentage of the wind's power and provide electricity at a lower price. According to the Center for Climate and Energy Solutions, efficiency has already increased dramatically in the past 30 years, with today's turbines producing 30 times as much power at a lower cost.

As wind technology improves, prices will continue to decrease. The price of wind power has already decreased significantly in recent years. According to a 2011 report prepared by the Lawrence Berkeley National Laboratory for the DOE, in the past 30 years the cost of wind energy production has decreased from 0.90 cents per kilowatt-hour to 0.06 cents per kilowatt-hour. This trend of decreasing prices is expected to continue as wind turbines become cheaper and

> "Wind power is steadily becoming one of the most cost-effective choices for electricity in the United States."[5]
>
> — Union of Concerned Scientists, a non-profit group that works toward improving the health of the environment.

more efficient. In the opinion of the Centre for Sustainable Energy, "There is significant potential for technological development in wind energy."[6]

Stability for Electricity Prices

Wind power installations do have a higher initial investment than fossil fuel power installations. After the cost of installation is paid off, however, a wind power plant is cheaper to operate than a fossil fuel one, and the price of the electricity produced remains stable. The initial investment to build a wind farm constitutes an estimated 75 to 80 percent of the total cost. The length of time it takes to pay off these start-up costs can vary considerably. However, the Natural Resources Defense Council estimates that in a location with good wind resources, it could take only 3 to 8 months. Once a wind farm has paid off its installation, costs are minimal because operating expenses are small, and the wind used to generate electricity is free. The Canadian Wind Energy Association discusses the way this creates long-term price stability for electricity produced by wind. It says, "Once a wind farm is built, the price of the electricity it produces is set and remains at that level for the entire life of the wind farm."[7]

In comparison, fossil fuel plants have substantial, and fluctuating, costs of fuel. For example, natural gas, which currently provides about 24 percent of the United States' electricity, is currently one of the cheapest ways to generate electricity. However, it has a history of price fluctuations. Although the price of wind-powered electricity generation will remain the same, says the Canadian Wind Energy Association, "There is no guarantee, for example, that natural gas will remain at today's low prices over the long term. Natural gas prices vary over time with changes in supply and demand—just a few years ago electricity from natural gas–fired projects was more expensive than electricity from wind."[8]

Not only do fossil fuel prices fluctuate, but they will gradually become more expensive over time as these fuels are depleted, and wind power will become increasingly affordable in comparison. Fossil fuels such as coal and natural gas are formed underground over millions of

Job Creation Enhances Affordability

In addition to being cost-competitive with other sources of electricity, wind power's affordability is enhanced by the number of jobs the industry creates. Overall, in 2010, an estimated 3.5 million renewable energy jobs were created worldwide—with the wind power industry playing a significant role in job creation. As this chart shows, China was the leader in wind industry job creation, followed by Germany and the United States. Experts expect job creation to increase as the industry grows.

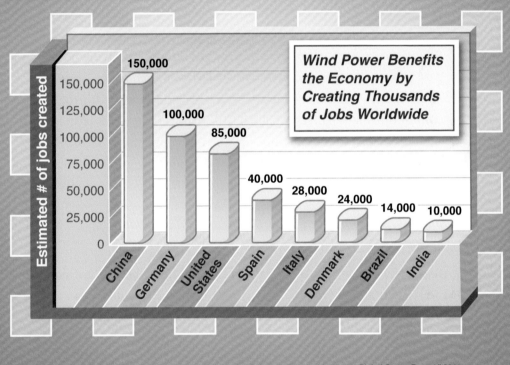

Wind Power Benefits the Economy by Creating Thousands of Jobs Worldwide

Estimated # of jobs created

- China: 150,000
- Germany: 100,000
- United States: 85,000
- Spain: 40,000
- Italy: 28,000
- Denmark: 24,000
- Brazil: 14,000
- India: 10,000

Source: Renewable Energy Policy Network for the 21st Century, "Renewables 2011: Global Status Report," 2011. www.ren21.net.

years. While there is disagreement over exactly how much of these fuels are left, it is generally agreed that they will gradually become more scarce as society continues to extract them and burn them for energy. Author Chris Nelder says, "Oil, natural gas and coal are set to peak and go into decline within the next decade, and no technology can change that." He estimates, "Natural gas is likewise expected to peak some time around 2010–2020, and coal around 2020–2030."[9] As supplies diminish, prices for fossil fuels will increase. As a result, wind power will become an increasingly affordable alternative.

Avoiding the Environmental Costs of Fossil Fuels

A true comparison between the affordability of wind power and fossil fuels should also include the indirect costs incurred by each technology, and this comparison further enhances the affordability of wind relative to other fossil fuels. The DOE explains indirect costs:

> All generation sources impose indirect costs on society that are not paid for by generators and therefore not reflected in the direct costs of electricity. Comparing the true costs of wind power with the costs of other alternatives usually reveals that wind power is favorable. Costs that are usually not included in comparisons include the costs of air, water, and land pollution from generation as well as fuel extraction and transport; nuclear waste disposal; oil spill prevention and cleanup; exposure to physical or economic disruption of supply lines; and military intervention to ensure supply.[10]

In a 2011 report, the Harvard School of Public Health estimated that the health and environmental costs of fossil fuels could be as much as $523.3 billion a year. In contrast, wind power has minimal environmental costs because it does not pollute the air, water, or land, and there is no fuel to extract and transport. Many people believe that if these environmental costs were included in the market price of fuel then wind power would actually be less expensive than fossil fuel–generated power.

Economic Benefits

Wind power also has numerous economic benefits. It creates jobs and generates local income. New jobs are created in manufacturing the various turbine parts and in building and maintaining wind power plants. Income is also generated through taxes and other payments made for the use of the land on which the plants are located. Many of these new jobs and much of the income generated by wind benefit local economies. According to a 2010 report by the DOE, in 2009 the wind sector employed 85,000 workers and invested more than $17 billion in the US economy. The US Department of the Interior Bureau of Land Management reports that wind power provides Americans with 5 times more jobs per dollar invested than coal or nuclear power. As in the United States, Europe reports significant economic benefits from wind power. According to the European Wind Energy Association, the wind power industry employs approximately 190,000 people throughout Europe, and that number will rise to more than 462,000 by 2020. In contrast to the local economic benefits created by wind, most countries that use fossil fuels for electricity must import these fuels, meaning that they are investing money into the economies of other countries instead of their own.

> "There is significant potential for technological development in wind energy."[6]
>
> — Centre for Sustainable Energy, an organization in the United Kingdom that looks for answers to the problems of climate change and rising energy costs.

When assessing the affordability of wind power compared to other sources of electricity, it is important to consider more than just the price of producing a unit of electricity. Wind power's price stability, environmental benefits, and local economic benefits add to its overall affordability and make it a good financial choice for electricity generation. Not only is it affordable now, but in comparison to other sources of electricity, wind will become even more affordable in the future.

Wind Power Is Too Costly

"Wind is the most expensive power in America."

—Tom Powell, a consultant who has worked for wind power constructors in the United States.

Tom Powell, "Wind Power the Most Expensive," *Lewiston (ME) Sun Journal*, May 10, 2010. www.sunjournal.com.

Although the price of wind power has decreased in recent years, it is still too costly. Electricity costs are often compared using a levelized cost, which is calculated by taking the building and operating costs for the power plant over its assumed lifetime, then using that total to calculate the average costs per unit of electricity produced. In a 2010 report, the US Energy Information Administration (EIA) estimated levelized costs for electricity generation for plants beginning service in 2016. The administration estimated that the average levelized cost of conventional coal will be $94.8 per megawatt-hour (MWh), conventional natural gas will be just $66.1 per megawatt-hour (MWh), and wind will be $97 per megawatt-hour (MWh). Even small price differences such as these are important and mean that compared to fossil fuel–generated power, wind power is simply too costly for society.

Abundant Coal and Natural Gas Reserves

Philip Totaro, whose company, Totaro and Associates, works with renewable energy companies to develop new products and technologies, explains that for most people price is more important than any other factor in choosing a source of electricity. As a result, he says, "as long as natural gas and coal are abundant and cheap, wind and other renewables will have a tough time displacing them. While it might be nice to see everyone take an interest in clean energy production for the sake of environmental preservation, consumers ultimately tend to speak with their wallets."[11]

The United States has abundant reserves of both coal and natural gas that can supply its electricity for many years to come, at a much lower cost than wind power. According to the EIA, the United States has the largest coal reserves in the world, with 28 percent of the world's estimated total. The agency states that at current mining levels this coal will last 222 years. The United States also has significant natural gas reserves. The estimates of these reserves continue to increase over time as exploration and production technologies improve. For example, proven natural gas reserves increased by 11 percent in 2009. The EIA estimates that at the 2010 rate of consumption, there is enough natural gas to supply more than 100 years of use in the United States. The Natural Gas Council says, "[Natural gas] can boost our energy security, advance our economy, [and] benefit consumers pocket books." As a result of the United States utilizing its abundant natural gas resources, the council's 2012 statement notes, "The average U.S. household will enjoy an increase in annual disposable income of $926 over each of the next three years."[12]

> "From an economic point of view the use of wind and solar energy production is an enormous waste of resources."[13]
>
> — K. de Groot and C. le Pair, Dutch engineers.

The Cost of Backup Power

Like other sources of power, wind has indirect costs that should be included in discussions of affordability. One is the cost of installing backup systems for wind power plants. Because wind blows intermittently and at variable speeds, and there is currently no economically viable way of storing the power generated from it, a wind power installation must be accompanied by a backup power system, usually fired by coal or natural gas. The backup system provides power when wind conditions are not ideal, and it increases the overall cost of using wind power.

Wind speeds are divided into 7 classes, with 1 being the slowest and 7 the fastest. Turbines only operate within a limited range of speeds. In order to generate power economically, a wind in class 3—at least 13 miles an hour (21kph)—is needed. However, when the wind is too fast, the turbine must be shut down or it can become damaged. In the United States, the areas with the best wind resources are generally located far from population centers, for example in Montana, North Dakota, and Wyoming.

Dutch engineers K. de Groot and C. le Pair have analyzed Germany's wind power generation facilities in relation to the variability of the wind. In their analysis, they report that Germany currently generates about 7.5 percent of its electricity from wind. De Groot and Le Pair find that because variations in wind do not match demand for electricity, conventional power stations are used and adjusted to match demand levels in Germany. This irregular pattern of use reduces the efficiency of the power stations and requires the use of more fossil fuel per unit of electricity generated. Overall, the engineers conclude, "From an economic point of view the use of wind and solar energy production is an enormous waste of resources."[13]

In a 2012 analysis, energy expert Benjamin Zycher echoes the conclusion that the need for backup power makes wind power too costly. He estimates the total cost of wind power, including the cost of maintaining backup power, and finds that when backup costs are included, wind power is very expensive. According to Zycher, using estimates for 2016, onshore wind power will cost about $149 per megawatt-hour. However when the cost of backup power is added, the total is $517 per megawatt-hour. This is significantly higher than the estimated $80 to $110 cost of gas or coal power.

Expensive Power Grid Improvements

In addition to the cost of backup power stations, as wind power generation expands, improvements and additions must be made to the existing power grid in order to integrate that power. Such additional investments make it very expensive to use wind power. In the United States, areas where wind power has expanded have also had electricity rate increases to cover these expansion costs. For example, in January 2011 many Oregon consumers experienced a significant increase when Pacific Power raised its rates by 14.5

Wind Power Loses in Cost Comparisons

Compared to coal, natural gas, and nuclear power, which currently provide most of the world's electricity, wind power (whether offshore or onshore) is a more expensive method of producing electricity. This graph compares the costs of electricity generation for these sources. Costs are in dollars per megawatt-hour of electricity produced, a common way to measure electrical power. Estimates are for plants beginning service in 2016.

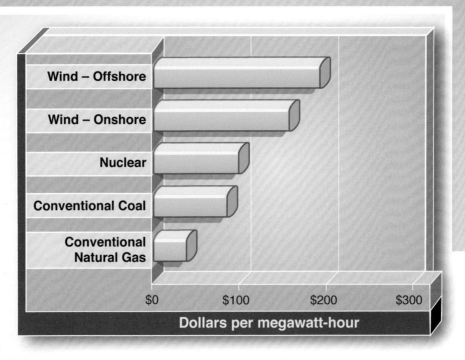

Source: Elliot H. Gue, "Sell Wind and Solar Energy Stocks," *Investing Daily*, April 14, 2010. www.investingdaily.com.

percent and Portland General Electric by 4.2 percent, in large part due to wind power expansion costs. The Citizens' Utility Board of Oregon reports that many Oregon residents are now struggling to pay their utility bills. In the first half of 2011, Oregon's three for-profit electric companies shut off power to more than 27,000 households for not paying their electric bills.

Many good wind resources are located far from where power is actually needed, so one cost associated with building wind farms is the construction of new transmission lines to carry the power from the farm to where it is used. Texas leads the United States in wind power production, getting about 8 percent of its electricity from wind. Yet in recent years, construction has slowed as existing transmission lines have become overwhelmed. In 2011 the *Texas Tribune* reported that almost $7 billion will be spent to build new transmission lines to carry wind power across Texas. However, the newspaper reported, this will raise electricity prices for consumers—an estimated $4 or $5 extra every month for years on every Texas electric bill.

Exaggerated Benefits

The so-called indirect economic benefits of wind power—especially job creation—have been exaggerated. Many of the parts and components for wind turbines are actually manufactured in other countries and only assembled in the United States. In 2010 Denmark's Vestas Wind Systems was the largest manufacturer of wind turbines in the world, and China's Sinovel Wind Group Company was the second largest. This means that many of the manufacturing jobs created by the wind industry are not local. What is more, construction jobs created during installation of wind farms are mostly temporary and often are filled by temporary workers bought in from other places. All of this significantly reduces the economic benefits to the local economy. And while wind power does create jobs, many people believe the overall effect is not beneficial because at the same time it takes jobs away from other industries.

> "The higher true cost of the electricity from wind is passed along to ordinary electric customers and taxpayers via electric bills and tax bills which means that people who bear the costs have less money to spend on other needs."[15]
>
> — Glenn Schleede, a wind energy expert and the author of many papers and reports on energy matters.

Zycher argues that when wind power and other renewable types of energy create jobs, they are not actually additional jobs for the economy as a whole. Instead, this job creation is simply a process of taking jobs from another part of the economy and putting them in wind power. He says, "An expanding renewable sector must be accompanied by a decline in other sectors . . . and creation of green jobs must be accompanied by destruction of jobs elsewhere."[14] As a result, he maintains, wind power actually harms the economy.

Energy expert Glenn Schleede believes that not only are many benefits exaggerated, but wind farms can actually harm local economies. He explains, "The higher true cost of the electricity from wind is passed along to ordinary electric customers and taxpayers via electric bills and tax bills which means that people who bear the costs have less money to spend on other needs (food, clothing, shelter, education, medical care—or hundreds of other things normally purchased in local stores), thus *reducing* the jobs associated with that spending and undermining local economies that would benefit from supplying these needs."[15]

Generating electricity from wind is not as simple as taking a free resource and converting it into inexpensive power. Not only does wind cost more in comparison to coal and natural gas, but backup stations and power grid improvements make it even more expensive. De Groot and Le Pair state, "Wind energy comes for free, but it does not follow that electricity generation using wind is also free."[16] Wind power is actually a costly choice that does not make good economic sense for society—now or in the future.

Chapter Two

How Does Wind Power Impact the Environment?

Wind Power Is Beneficial to the Environment

When compared to other means of electricity generation, wind power is beneficial to the environment. It can help reduce global warming and decrease the land, water, and air pollution currently caused by fossil fuel power generation. Wind power can also help society conserve precious water resources because it uses much less water than other types of power generation. Although some birds and bats are killed in wind turbines, studies show that turbines kill far fewer birds than many other types of human activity, including fossil fuel power generation.

The Debate

Wind Power Does Not Benefit the Environment

Wind power does not benefit the environment. When the wind stops blowing, backup power plants burn fossil fuels to keep electricity flowing, and the resulting pollutants contribute to global warming. Additionally, wind farms that might have upward of 100 massive turbines are a blight on the landscape. Aside from ruining scenic views, wind farms have displaced birds and bats from their habitats and killed hundreds as a result of collisions. Further, people who live near these wind farms suffer health effects from the noise and vibrations generated by the turbines.

Wind Power Is Beneficial to the Environment

"Wind power offers substantial public health, economic, and environmental benefits. It produces no air or water pollution, global warming emissions, or waste products, and saves water."

— Union of Concerned Scientists, a science-based nonprofit group that works to improve the health of the environment.

Union of Concerned Scientists, "Tapping Into Wind Power," April 2011. www.ucsusa.org.

Wind power is a clean, efficient, nonpolluting form of energy. It can help the world reduce greenhouse gas emissions and, as a result, reduce global warming. Burning fossil fuels creates carbon dioxide gas, which is widely believed to contribute to global warming. Experts believe that as the world's temperature rises through global warming there will be many harmful impacts on the environment and human health, including extreme weather events, reduced food production, and species extinctions. According to a 2011 report by the US Environmental Protection Agency, electricity generation accounted for 41 percent of the carbon dioxide emissions from fossil fuel combustion in 2009. In comparison, wind power causes only minor emissions from the manufacturing and transportation of turbines, and none from electricity generation. The DOE estimates that the United States emits about 6.6 billion tons (6 billion metric tons) of carbon dioxide every year. If the United States met the goal of producing 20 percent of its electricity through wind power by 2030, the DOE calculates that emissions in the electric sector would be reduced by 909 million tons (825 million metric tons). The DOE says, "The threat of climate change and the growing attention paid to it are helping to position wind power as an increasingly attractive option for new power generation."[17]

Electricity Without Pollution

Not only does wind power produce no greenhouse gas emissions, but it does not create the myriad types of pollution that other sources of electricity do. The mining and burning of fossil fuels to generate electricity pollutes the air, land, and water, and threatens the health of all living things. Maine's public health director, Dora Anne Mills, argues that using wind instead of fossil fuels to generate power would prevent much environmental harm. She says, "Generating energy from wind turbines means less energy generated from . . . oil and coal, both being major contributors to global warming, pollution, and resulting diseases and deaths due to heart disease, cancer, asthma, and other lung diseases."[18]

Coal power plants generate the majority of US electricity, and they cause significant pollution. These plants create air pollutants that cause or aggravate respiratory disease; cause acid rain, which harms the environment; and emit mercury, which in high concentrations can cause birth defects or brain damage in young children. According to a 2011 report by the American Lung Association, air pollution from power plants—primarily those burning coal—kills approximately 13,000 people every year in the United States. Another study in the *Annals of the New York Academy of Sciences* in 2011 found that the health costs of cancer, lung disease, and other illnesses connected to air pollutants is more than $185 billion each year. Coal mining can also result in deforestation and habitat destruction and release toxic amounts of heavy metals into the soil and water.

A Better Use of the Environment

Using wind to generate electricity does not bring about the same widespread habitat destruction and pollution as the use of coal and other fossil fuels. While wind farms do require large areas of land, much of that land can simultaneously be used for other purposes, for example farms, highways, and hiking trails. The Union of Concerned Scientists explains that the actual turbines take up a very small area, with large areas of land left as spacing between them. It says, "Turbines and related infrastructure occupy just 2 to 5 percent of that area, leaving at least 95 percent of the land free for other uses."[19]

Wind power generation also uses far less water than fossil fuel power generation, so it helps the environment by conserving water, a resource that is scare in many places. Almost half of the water withdrawn from the water supply system in the United States is for electricity generation. Although most of this water is recycled back into the system, according to the DOE, approximately 2 to 3 percent is lost, which adds up to about 1.6 trillion gallons (6 trillion L) of water being used for power generation every year. In contrast, wind power uses almost no water. Steve Sawyer, secretary-general of the Global Wind Energy Council, says, "Wind power can make a considerable contribution to conserving the world's valuable water resources. Unlike most other power sources, which consume huge amounts of water that could be used much more productively for human consumption and agriculture, wind power generation does not use any water."[20]

> "The threat of climate change and the growing attention paid to it are helping to position wind power as an increasingly attractive option for new power generation."[17]
>
> —The DOE, the US government agency that works to advance energy technology in the United States.

Safe for Birds

Despite claims to the contrary, wind power is actually safer for bird populations than other types of power plants. Researcher Benjamin K. Sovacool points out that although wind turbines receive more attention in terms of bird deaths, nuclear and fossil fuel power stations are actually a far greater threat to birds. He investigated bird deaths from wind power generation in the United States and Europe, then compared this to bird deaths caused by nuclear power plants and fossil fuel power stations. He found that wind power plants actually kill far fewer birds than the other types of power generation. In his 2009 report, he wrote that according to his analysis, wind farms kill about 7,000 birds a year, but fuel power plants kill 14.5 million. These plants cause bird deaths through collisions with plant operating equipment and transmission cables and through environmental destruction and pollution from the mining and burning of coal.

Wind Power Avoids Environmental Harms of Coal

Using wind to generate electricity has a minimal environmental impact compared to the production of electricity through fossil fuels such as coal and natural gas. These diagrams compare the effects of a typical coal power plant with a wind power facility during the electricity production process. They show that producing electricity from coal requires large amounts of water and coal, while wind power requires only wind, a free and unlimited resource. In addition, a coal power plant emits a significant amount of pollution while generating electricity, while a wind power plant facility does not.

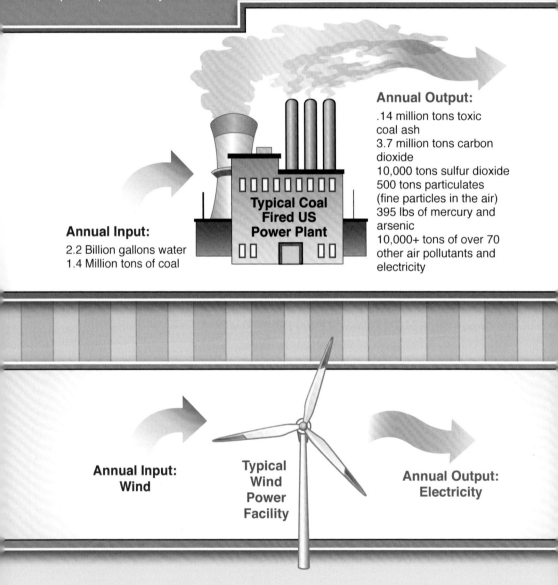

Typical Coal Fired US Power Plant

Annual Input:
2.2 Billion gallons water
1.4 Million tons of coal

Annual Output:
.14 million tons toxic coal ash
3.7 million tons carbon dioxide
10,000 tons sulfur dioxide
500 tons particulates (fine particles in the air)
395 lbs of mercury and arsenic
10,000+ tons of over 70 other air pollutants and electricity

Typical Wind Power Facility

Annual Input:
Wind

Annual Output:
Electricity

Source: Robert E. Beauford, "How Green Is Green? Comparing the Environmental Impacts of Wind Power and Coal Power," *Rare Earth Elements*, April 16, 2011. www.rareearthelements.us.

What is more, the percentage of birds killed by wind turbines is miniscule when compared with bird deaths resulting from other human activities and human lifestyles. According to a 2011 report by CLF Ventures and the Massachusetts Clean Energy Center, between 500 million and 1 billion birds are killed each year in the United States in connection with human activity. Of these, only 0.003 percent are estimated to be due to wind turbines, whereas 82 percent are killed by cats and collisions with buildings and power lines.

While some early wind farms did kill significant numbers of birds, these deaths were the result of a lack of research and understanding about the effect of turbines on birds. Ongoing research and improved turbine design now allow wind farms to be constructed and operated with a minimal impact on birds. The Centre for Sustainable Energy says, "Wind turbines have earned a bad reputation thanks to several large installations built in the early 1980s. . . . [For example,] early wind turbines were sited with very little consideration for the indigenous raptor populations in the [Altamont Pass Wind Resource Area in California]." However, says the organization, "This effect is not observed to such a degree in similar wind farms sited in the USA leading to the conclusion that poor planning and outmoded turbine design is largely responsible."[21]

No Proven Harms to Human Health

The effects of wind power generation on human health have also been studied, with the same general conclusion: Other forms of energy generation are more harmful than wind turbines. The most frequent concern has to do with turbine noise. Some people who live close to wind turbines report health problems that they believe have been caused by the turbines. However, researchers contend that these health complaints are exaggerated and unproven, and that fossil fuel power generation has far greater health harms.

In a report prepared for the American Wind Energy Association and the Canadian Wind Energy Association, medical experts reviewed existing research and concluded that the sound and vibrations of wind turbines do not present any risk to human health. Instead, they believe that the reported symptoms of turbine exposure are simply that a small group of people experience annoyance and stress in reaction to the turbine noise. The researchers conclude, "Associated stress from annoyance, exacerbated by the rhetoric, fears, and negative publicity generated by the wind turbine controversy, may contribute to the reported symptoms described by some people living near rural wind turbines."[22]

Research from Netherlands supports the theory that some of the health problems associated with wind turbines may be related to mental attitude rather than any actual physical harm. Researchers found that people who are annoyed by the presence of wind turbines were more likely to report health problems. Of those people studied, the ones who found the wind turbines ugly were more likely to notice the noise they generated and to be annoyed by it. Attorney Trey Cox, who represented a wind farm developer in a Texas lawsuit says, "I think that when people don't like the wind turbine, they become bigger, they become louder and they become uglier in their minds."[23]

> "Wind power can make a considerable contribution to conserving the world's valuable water resources."[20]
>
> —Steve Sawyer, secretary-general of the Global Wind Energy Council, a wind energy organization.

Minimizing Environmental Costs

No matter how society generates its electricity, there is likely to be some environmental harm. Energy expert Robert Bryce argues, "Every source of energy production takes a toll on the environment." In his opinion, "The goal should be to minimize these costs."[24] Society requires large amounts of electricity, and as the world's population grows each year, so will electricity needs. Using fossil fuels to generate that electricity has proved extremely harmful to the environment. Instead, society could minimize environmental harms and produce the electricity it needs by using wind power.

Wind Power Does Not Benefit the Environment

"Wind turbines do have negative impacts on the environment."

—US Energy Information Administration, the statistical and analytical agency within the DOE.

US Energy Information Administration, "Renewable Wind," *Energy Kids*, 2010. www.eia.gov.

Wind power is not the environmental panacea its supporters say it is. As a solution to global warming, wind power is a poor alternative. Claims that wind power can substantially reduce carbon dioxide emissions, and thus reduce global warming, cannot be substantiated. Explains energy expert Robert Bryce, "Given the hype about wind power, it would be logical to assume that wind-power advocates have multiple studies on their shelves to prove that wind power cuts carbon dioxide emissions. The problem: they don't have a single study to support their claim."[25] Instead, says Bryce, reports are simply based on models and projections about what carbon dioxide reductions *might* be.

In a 2011 report, Bryce argues that if the United States could achieve its goal of obtaining 20 percent of its electricity from wind by 2030, it would only reduce global carbon dioxide emissions by about 2 percent. Not only is this reduction very small, argues Bryce, but it would come at a high cost to society because wind power is more expensive than coal or natural gas. He estimates that in such a scenario, the cost of residential electricity in coal-dependent regions of the country would increase by as much as 48 percent.

Wind Power Pollutes

What is more, wind power is not a clean, nonpolluting energy source. Wind power only works when the wind blows. When the winds die down, people and businesses still need power. The only way to ensure uninterrupted energy flow is to have a backup power source—usually fossil fuel generators, which contribute significantly to environmental pollution. The *National Review* states that the use of backup generators cancels out the environmental benefits of wind power. It says, "To be available at a moment's notice, gas generators need to run on standby—burning fuel but producing no power—24 hours a day, and the emissions from this more than outweigh the savings from using wind."[26] In addition to always running, fossil fuel power plants that are used for backup must continually change their level of power generation, making sudden increases and decreases to match the level of the wind. Known as cycling, this process also results in increased emissions. A 2010 report sponsored by the Independent Petroleum Association of Mountain States examined the use of wind energy in Colorado. It concluded that because of the erratic and unpredictable nature of the wind, the coal power plants used as backup must also be used erratically, and as a result are an environmental hazard. The researchers state, "Cycling coal plants to accommodate wind generation makes the plants operate inefficiently, which drives up emissions. Moreover, when they are not operated consistently at their designed temperatures, the variability causes problems with the way they interact with their associated emission control technologies."[27] They find that overall, emissions of sulfur dioxide, carbon dioxide, and nitrous oxide—gases that contribute to air pollution and global warming—have actually increased in the Colorado power plants studied, due to the use of wind energy.

> "It would be logical to assume that wind-power advocates have multiple studies on their shelves to prove that wind power cuts carbon dioxide emissions. The problem: they don't have a single study to support their claim."[25]
>
> —Robert Bryce, the author of three books and numerous articles about energy.

Detrimental to the Landscape

Wind farms are usually located in rural areas where there is wind and space for the turbines. Some wind turbines are 200 or 300 feet (61m or 91m) tall, and a large wind farm with hundreds of turbines can ruin peaceful settings and scenic views. Environmentalist Kelvin Duncan maintains that both the appearance and movement of these turbines detracts from the beauty of a landscape. He says, "I think wind turbines are immensely ugly and intrusive. And what's more—they move in a mechanical, vaguely threatening manner. The appearance of the landscape is important, and they destroy landscapes."[28]

Most people do not want turbines located close to their own homes because of how large and intrusive they are. For example, Massachusetts resident Chad Pepin disputed a proposed wind turbine next to his property, insisting: "Monstrous moving towers don't belong in MY BACK YARD. They don't belong in ANYONE's back yard."[29]

Harmful to Birds

Wind turbines also have a harmful effect on wildlife, particularly birds and bats. According to the US Fish and Wildlife Service, about 440,000 birds are killed at US wind farms each year. California's Altamont Pass Wind Resource Area, which has about 5,000 turbines, is an example of a wind farm that has been particularly harmful to birds. It is located in a region with one of the highest densities of nesting golden eagles in the United States. Research shows that an average of 67 eagles have died there every year for the past 30 years due to the turbines, a number greater than can be replaced by breeding among existing eagles.

The United States' stated goal of generating 20 percent of its electricity through wind power by 2030 could devastate US bird populations. The American Bird Conservancy says that wind turbines will kill at least 1 million birds a year, and possibly many more, once that goal is reached. Mike Parr, president of the conservancy, insists that there should be more research before additional wind farms are constructed. He says, "We are plunging head-long into wind power, but so far, very few studies have been conducted that show what scale of impact it will really have on birds."[30]

A Cause of Bat Deaths

Even less well understood is the impact of wind turbines on bats. Some experts believe that turbines may be more damaging to bats because there is a smaller number of species involved. Like birds, bats are killed in collisions with turbines, but they are also harmed by flying close to the turbines, where the sudden drop in air pressure caused by the swinging rotors can cause their blood vessels and organs to rupture. In 2011 a study by the University of Wisconsin–Madison was published showing that experts may be underestimating the number of bats being killed by turbines. Researchers examined bats that died near wind turbines and reported that many had injuries such as broken bones, soft tissue damage, and ruptured eardrums. These types of injuries are not instantly fatal to bats, so researchers believe that many bat deaths are not being discovered because bats are leaving the vicinity of the turbines before they die. Some people believe that wind turbines may be particularly harmful to bats because bats live a long time and have low reproductive rates. This means that bat populations may have difficulty recovering from large numbers of deaths due to turbines.

Declining bat populations could also have a negative impact on farmers. Bats eat large quantities of bugs that would otherwise ruin crops, so they save farmers millions of dollars every year. According to one expert, whose comments appeared in the *Pittsburgh Post-Gazette* in 2011, bats consume as many as 500 insects in one hour, or nearly 3,000 insects in one night. The newspaper says, "If one turbine kills 25 bats in a year, that means one turbine accounted for about 17 million uneaten bugs in 2010."[31] These uneaten bugs could cost farmers a lot of money in crop damage and reduced harvests.

Harmful to Society

Finally, wind power—even on a small scale—can harm human health. Experts generally agree that certain types of noise can cause health problems. According to the World Health Organization, environmental noise can be a threat to public health. Wind turbines create noise through the vibration of the gearbox and the generator, and also from the rotation of the blades. In addition, the rotating blades create an effect of moving shadows known as "shadow flicker." Some people who live near turbines complain that the noise and the constant shadow flicker have been

Wind Power Is Killing Bats

Significant numbers of bats are dying at some US wind power plants as a result of collisions and air pressure changes near the turbines. Research on bat deaths is limited, however one 2010 study found bat deaths to be a bigger problem at some wind facilities than others. According to the study, bat fatalities were highest at the wind facility at Buffalo Mountain, Tennessee. Bat fatalities were also relatively high at wind power plants in West Virginia and Wisconsin. Many researchers believe that bat fatalities are being underestimated and that further study is needed.

Summary of bat mortality rates at various wind energy facilities

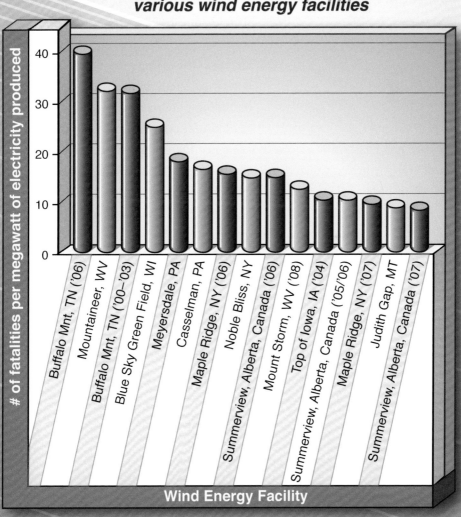

Source: National Wind Coordinating Collaborative, "Wind Turbine Interactions with Birds, Bats, and Their Habitats." Spring 2010. www.nationalwind.org.

harmful to their health. New York doctor Nina Pierpont has researched these health problems and published her findings. She calls the problem "wind turbine syndrome." The syndrome has symptoms of sleeplessness, headaches, dizziness, depression, and nausea.

There are many reports of these symptoms from people who live near wind farms. For example, Hal and Judy Graham live in upstate New York and have two wind turbines near their house—one is 1,000 feet (305m) away and the other is 2,000 feet (610m) away. Says Hal, "We can't sleep. We can't watch TV. This has been a disaster for us and our neighbors."[32] Jane Davis lives near Linconshire, Great Britain. She reports that after the construction of eight turbines within 3,000 feet (914m) of her home, she and her family suffered many health problems including pneumonia, hearing loss, and sleep problems. She says, "My daughter Emily and I have not slept at home since December 2006, a fact which although ensuring we have good sleep, causes us considerable distress."[33]

> "Wind turbines are immensely ugly and intrusive. . . . They destroy landscapes."[28]
>
> —Kelvin Duncan, an environmentalist and former dean of science at the University of Canterbury in Christchurch, New Zealand.

Wind power does not affect the environment in the same way as fossil fuels. However, that does not mean it is beneficial to the environment. Wind power creates its own environmental harms. It causes pollution, kills birds and bats, and causes numerous health problems in people who spend long periods of time near turbines. Expanding wind power in the future will only exacerbate these problems.

Chapter Three

Can Wind Power Ever Replace Fossil Fuels?

Wind Power Can Replace Fossil Fuels

Fossil fuels will become increasingly scarce and expensive, and the world will need to replace them with another source of power. Wind power is abundant and nonpolluting, and it will be a big part of the solution. Although most existing turbines are located onshore, offshore wind turbines will become increasingly widespread in the future. They will help society generate the large amounts of electricity it needs, while avoiding many of the problems caused by onshore turbines. Integrating large amounts of wind power into the existing grid will be challenging but can definitely be accomplished.

The Debate

Wind Power Is Not a Viable Replacement for Fossil Fuels

Fossil fuels are a cheap and reliable power source, and wind will probably never be a reliable replacement. There are some fundamental conflicts between the nature of the wind and the nature of electricity generation today that make it impossible to rely heavily on wind power. The difficulty of integrating wind into the power grid and of getting it to where it is needed are also problems. Finally, even with widespread wind power, fossil fuels will not be eliminated because they will be required as a backup power source.

Wind Power Can Replace Fossil Fuels

> "Wind is the most likely candidate for driving the long-awaited shift [from fossil fuels] toward renewable energy."

—Daniel Weiss, a senior fellow at the Center for American Progress.

Quoted in Erik Heinrich, "Breezing In," *Time*, June 20, 2011. www.time.com.

Wind power is a good replacement for fossil fuels because in the future, fossil fuels will become increasingly limited and expensive, whereas wind will always be free and will never run out. The world has a finite reserve of fossil fuels that is steadily being depleted and will one day be gone. These fuels are an easy source of power now because they are still relatively abundant, but as society continues to use large amounts of fossil fuels for its energy needs, they will become increasingly scarce and expensive. Experts disagree about when this scarcity will start to have a serious impact on society, but most agree that it is inevitable. David J.C. MacKay, professor of physics at the University of Cambridge and author of *Sustainable Energy—Without the Hot Air*, argues that although society depends on fossil fuels for electricity now, it cannot do so forever. He says, "This era of easily accessible fossil fuels is likely to be but a brief blip in the history of humanity. The peaks of oil and gas production are expected to be reached within the next 50 years, and coal production is likely to peak about the end of the century."[34] Eventually, fossil fuels will not be an easy source of energy and society will need to use something else. Wind is a good alternative.

Offshore Turbines

In order to replace fossil fuels, wind power will need to generate large quantities of electricity, and this can be accomplished by using offshore wind turbines in addition to onshore ones. These turbines, located in bodies of water such as lakes or coastal seas, have significant potential to provide large amounts of electricity while avoiding many of the problems of onshore turbines. Offshore wind farms are more expensive to construct, but they also provide greater amounts of electricity than onshore ones. One advantage of offshore turbines is that the wind blows harder and more reliably offshore. In addition, whereas onshore wind frequently blows hardest at night, offshore wind is often strong in the afternoon, when people use a lot of electricity. Finally, locating turbines offshore reduces conflicts over noise because the turbines are located far away from where people live. In the opinion of Danian Zheng and Sumit Bose of the General Electric Company, this technology has great potential. They say, "Offshore wind presents a tremendous opportunity for the United States and Europe."[35]

> "This era of easily accessible fossil fuels is likely to be but a brief blip in the history of humanity." [34]
>
> —David J.C. MacKay, professor of physics at the University of Cambridge and author of *Sustainable Energy—Without the Hot Air*.

Offshore wind technology is still relatively new, but offshore wind farms are increasing as various countries recognize their potential. In 1991 Denmark built the world's first offshore wind turbine. Today offshore wind farms are located in numerous other countries, including the United Kingdom, Belgium, Netherlands, Sweden, and China. According to the Center for Climate and Energy Solutions, the United Kingdom leads the world in offshore wind power capacity, with 1,341 megawatts, followed by China, with 854 megawatts.

After years of regulatory debate, the United States approved the construction of its first offshore wind farm in 2010. Cape Wind will be located in Nantucket Sound, off the East Coast of the United States. It will comprise 130 turbines, which its developer hopes will provide about 75 percent of the electricity used by the residents of Cape Cod and the islands of Martha's Vineyard and Nantucket. Cape Wind says, "By harnessing our local wind resources, we can contribute to reducing our dependence on imported energy. Cape Wind will provide clean, renewable energy capable of displacing 113 million gallons of oil per year."[36] Other US wind farms have been proposed, but none has yet been approved.

Global Wind Power Capacity Grows Yearly

Wind power is becoming increasingly viable as a replacement for fossil fuels. This graph shows cumulative installed wind capacity worldwide, or the amount of power the world is capable of generating with existing wind turbines. It reveals that wind power's global capacity is increasing greatly every year, meaning that wind power is capable of providing increasing amounts of the world's electricity. Capacity is measured in megawatts, a common way of measuring electricity.

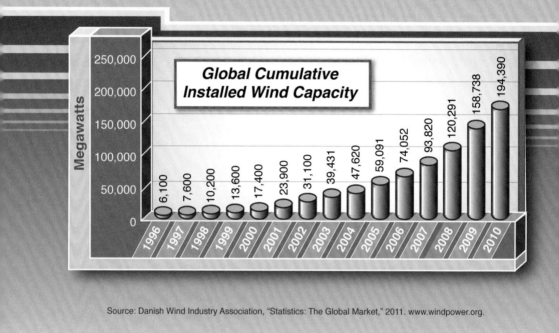

Global Cumulative Installed Wind Capacity

Year	Megawatts
1996	6,100
1997	7,600
1998	10,200
1999	13,600
2000	17,400
2001	23,900
2002	31,100
2003	39,431
2004	47,620
2005	59,091
2006	74,052
2007	93,820
2008	120,291
2009	158,738
2010	194,390

Source: Danish Wind Industry Association, "Statistics: The Global Market," 2011. www.windpower.org.

In a 2011 report the Departments of Energy and the Interior stress that offshore wind farms should be an important part of wind power expansion in the United States.

Great Potential

A variety of forms of power generation will be needed if society is to replace fossil fuels, and wind power has the potential to become one of the most important replacements. The Intergovernmental Panel on Climate Change estimates that globally, wind energy is only at 0.4 percent of its estimated potential for electricity production. Researchers state, "Wind electricity penetration levels that approach or exceed 10% of global electricity supply by 2030 are feasible." They state: "The scenarios further suggest that even more ambitious policies and/or technology improvements may allow wind energy to reach or exceed 20% of global electricity supply by 2050."[37]

Some US states and a number of countries have already managed to successfully use wind power to replace a significant amount of their fossil fuel–generated power. Iowa and Texas obtain a significant amount of their electricity from wind power and Germany, Spain, Portugal, Denmark, and Ireland all get more than 10 percent of their electricity from the wind.

Integrating Wind Power into the Grid

Although there are challenges, large amounts of wind power can be successfully integrated into the power grid. Wind power is more difficult to integrate than fossil fuel power because of its variable nature. However, power operators are becoming increasingly skilled at handling this variability. In fact, while wind variability is a new type of challenge for power operators, they already deal with significant variability in fossil-fueled power too. Explains the American Wind Energy Association (AWEA), "Factories turning large electrical equipment on and off and millions of people changing their use of air conditioning and electric heating as the weather changes cause large and often unpredictable changes in the demand for electricity. Similarly, large changes in electrical supply occur fairly frequently when large conventional power plants experience sudden outages dues to mechanical or electrical failures." According to AWEA, "While the casual observer does not see what is going on behind the scenes to keep the lights on, it is actually a very sophisticated balancing act."[38]

Operators can learn to handle variations in wind power the same way they have learned to balance variations in fossil fuel–generated power. In addition, it is hoped that in the future researchers will develop storage options for wind energy such as batteries. This will allow power operators to store some wind power for use when the wind does stop blowing.

Peter Musgrove, a wind power expert who has been involved in many aspects of the development of wind power in Great Britain, explains that turbine operators can also compensate for wind variability by forecasting when and where it will occur. In this way, he says, operators can anticipate the variability of wind in different regions and compensate for a lack of wind in one area by using wind from another. Musgrove explains that although weather may change quickly in one part of a country, overall a weather system is large and relatively predictable. He uses the example of wind power in Great Britain to illustrate his point: "The air over just a single region of England weighs many billions of [tons]; and when it is moving—like a supertanker, only more so—it cannot suddenly stop. Weather systems in fact take many hours to cross a country such as the UK."[39] As a result, Musgrove argues, the problem of variability from individual wind turbines can be solved by combining the power output of turbines across the country, which changes relatively slowly.

Challenges for the Future

Another challenge to using large amounts of wind power is that new transmission systems must be built to accommodate that power. However, in the United States, experts believe new transmission lines will be necessary regardless of the power source used. This is because the United States—like many other countries, including China and India—is using increasing amounts of electricity every year. As the country generates increasing amounts of power to meet demand, new transmission lines will be needed to carry it to customers. Says the DOE, "If electric loads keep growing as expected, however, extensive new transmission [systems] will be required." According to the DOE, "Over the coming decades, this will be true regardless of the power sources that dominate, whether they are fossil fuels, wind, hydropower, or others."[40]

Because the United States relies so heavily on fossil fuels, its actions will be important in determining whether or not the world can replace fossil fuels with wind power. The country has only about 5 percent of

> "Offshore wind presents a tremendous opportunity for the United States and Europe."[35]
>
> —Danian Zheng and Sumit Bose, of the General Electric Company; Zheng is in Infrastructure Energy and Bose is in the Global Research Center.

the world's population but is the world's second-largest user of coal, after China. Most of that coal is used for electricity generation. So if the United States can replace some of its coal consumption with wind power, it will have a significant impact on the world's progress toward replacing fossil fuels. Because of this heavy reliance on fossil fuels, replacing them will be challenging and costly. As energy expert Robert Bryce explains, "Over the past century or so, the United States has built a $14-trillion-per-year economy that's based almost entirely on cheap hydrocarbons."[41] However, despite the challenges, it is believed that it is feasible for the United States, and the world, to replace a significant amount of fossil fuels consumed. Wind power will play an important part in achieving this goal.

Wind Power Is Not a Viable Replacement for Fossil Fuels

"Thanks to its variability and intermittency, wind power does not, and cannot, displace [fossil fuel] power plants, it only adds to them."

—Robert Bryce, the author of four books and numerous articles about energy.

Robert Bryce, *Power Hungry: The Myths of "Green" Energy and the Real Fuels of the Future.* New York: PublicAffairs, 2010, p. 99.

Fossil fuels provide large amounts of electricity without the need for large power stations. Wind power is simply not a viable replacement for this technology. It requires too much land to generate significant amounts of electricity. It would be impossible to cover enough area with turbines to generate the electricity society needs. In a 2011 *New York Times* article, energy expert Robert Bryce provided an example using California, which has a mandate to provide one-third of its electricity from renewable sources by 2020. He says that to get even half that—about one-sixth of its electricity—from wind power would require an area of land equivalent to 70 Manhattan-sized cities. Critic Bjorn Lomborg argues, "There is no affordable alternative to fossil fuel. Current green technology is so inefficient that—to take just one example—if we were serious about wind power, we would have to blanket most countries with wind turbines to generate enough energy."[42] In addition, even if it were possible to blanket most countries with wind turbines, there would still not be enough electricity to replace fossil fuels because not all areas have suitable wind resources. For example, forested areas and urban areas generally have poor wind exposure.

Increasing Population

Not only can wind power not match the level of power used today, but global energy demand is increasing rapidly and will leave wind power even further behind. The world population increases every year. According to the US Census Bureau, the current world population is about 7 billion. By 2040 it is expected to reach almost 9 billion. This increase will mean a dramatic rise in electricity needs. In a 2011 report, the EIA estimated that between 2008 and 2035, global energy demand will increase by 53 percent.

Wind power and other renewable energy sources are increasing rapidly too, but by 2035 the EIA projects that all renewable energy sources put together will only make up 15 percent of total energy production. With this kind of growth rate, it will take a very long time for wind power to even make up a significant percentage of current demand, let alone meet new demand. Richard Stieglitz, an expert on nuclear power, and journalist Rick Docksai argue that growing energy demand is already unsustainable. They state, "Communities around the globe already feel the pinch of rising energy prices and electricity shortages, so a potent new source of energy will clearly be needed. Solar, wind, geothermal, and other renewable resources will help, but few analysts see them as sufficient in themselves."[43]

Simply trying to replace fossil fuels with wind power is the wrong way to approach the problem of meeting society's energy needs. A better solution is to focus on energy conservation. The organization Stop Ill Wind argues that society is addicted to energy use in the same way that a smoker is addicted to nicotine. It says that trying to find a new source for that energy, such as wind, is avoiding the real problem. It likens such an approach to a heavy smoker who tries to cut down on smoking by switching to light cigarettes rather than smoking fewer cigarettes altogether. It says, "Industrial wind plants, in their current incarnation, are to the reduction of dependence on fossil fuels as the smoker who seeks to mitigate the dangers of smoking by switching to three packs of Marlboro [Lights]."[44] Instead of searching for other ways to satisfy its energy addiction, society should look for ways to reduce its energy use.

> "If we were serious about wind power, we would have to blanket most countries with wind turbines to generate enough energy."[42]
>
> —Bjorn Lomborg, a Danish environmentalist and author of a number of books and articles on environmental topics.

Fossil Fuels Will Still Dominate Electricity Production in the Future

According to projections by the US government, wind power and other renewable forms of energy will provide increasing amounts of the world's electricity in the future, but they will not replace fossil fuels. This graph shows the fuels that comprise world electricity generation now, and provides estimates for the future. It shows that by 2035, renewable sources of energy are expected to comprise a greater percentage of the total, however fossil fuels will still make up more than half of total electricity generation. Estimates are in trillion kilowatt-hours, a common measurement of energy.

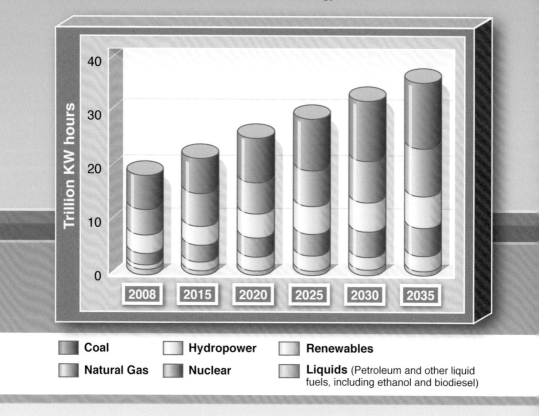

Legend:
- Coal
- Hydropower
- Renewables
- Natural Gas
- Nuclear
- Liquids (Petroleum and other liquid fuels, including ethanol and biodiesel)

Source: US Energy Information Administration, "International Energy Outlook 2011," September 19, 2011. www.eia.gov.

An Unpredictable Source of Power

Another major problem with using wind power to supply large amounts of electricity is the intermittency of wind, which makes it very difficult to provide the continual supply of power that people and businesses need.

Fossil fuels provide a predictable level of electricity, and they provide it according to consumer demand. In comparison, wind power is unpredictable and frequently does not match consumer demand. Furthermore, sometimes the wind stops blowing altogether. Ed Hiserodt, a power generation technology expert, explains the way conventional power generation is based on meeting fluctuating consumer demand. He says:

> Power suppliers must contend with fluctuating power demands, both daily and seasonal changes. At 5:30 a.m. alarm clocks start ringing, coffee pots start up, along with hot water heaters for showers. Restaurants fire up toasters, and factories come up to speed for a day of production. Grid operators expect this to happen and, based on hour of day, time of year, and day of week, bring on additional generating assets, such as small coal plants. . . . Can wind power be scheduled by operators to follow the daily variations in demand? Hardly.[45]

Wind power is simply not compatible with the daily needs of modern societies.

Grid Problems

In addition to supplying electricity on demand, an electrical power grid needs to remain at a stable level, without big fluctuations. Wind power fluctuates constantly. This means it is not a good source of electricity because it destabilizes the grid, forcing operators to work constantly to counteract its fluctuations. Environmentalist Jon Boone explains that because of its unpredictable nature, wind must be stabilized with another, more predictable source of electricity. He says:

> Integrating wind's skittering energy levels within the mix of reliably steady conventional generation is a daunting challenge. Achieving this integration insures there will be increased financial and environmental costs. The now-you-see-it, now-you-don't nature of wind necessitates that it be continuously accompanied by reliable compensatory generation in order to maintain a steady power flow matched to demand. Consequently, wind can only be one ingredient in a larger fuel mix.[46]

The proliferation of wind farms in the northwestern United States has led to many recent power grid problems caused by wind fluctuation. Grid operators cannot integrate so much fluctuating power. According to one article on an Oregon news website, "The pace and geographic concentration of wind development, coupled with wild swings in its output are overwhelming the region's grid and outstripping its ability to use the power or send it elsewhere."[47] This inability to integrate the power means that it is being wasted. In some cases, wind farm operators have been forced to shut down their turbines when the wind is strong because grid operators cannot integrate so much power at once.

Not Replacing Fossil Fuels

Even if wind could be used to provide a significant percentage of the world's electricity, it would not replace fossil fuels, because in most cases fossil fuel power stations are still needed for backup power. There is currently no cost-effective way to store excess wind energy, so when the wind slows or stops, a backup source of power is required to maintain the electricity supply to consumers. In most cases, that is a fossil fuel power station, and it must run continuously to supply power when it is needed. Keeping these stations running means that there is no significant reduction in the amount of fossil fuel used.

> "Integrating wind's skittering energy levels within the mix of reliably steady conventional generation is a daunting challenge. . . . Consequently, wind can only be one ingredient in a larger fuel mix."[46]
>
> —Jon Boone, an environmentalist and a founder of the North American Bluebird Society.

Bryce examines the case of Denmark, a world leader in wind power, to see whether increased wind power decreases the use of fossil fuel. He says that between 1999 and 2007, the amount of energy produced by wind turbines in Denmark increased by 136 percent, but coal consumption did not change at all. In his opinion, "The basic problem with Denmark's wind-power sector is the same as it is everywhere else: It must be backed up by conventional sources of generation."[48] This conventional generation means that fossil fuel use remains largely the same even when wind power is used.

At present, there is no feasible solution to the storage problem and this severely limits the ability of wind power to replace fossil fuels. *Forbes* magazine contributor Gene Marcial insists, "There is no replacing fossil fuel until a way is found to store renewable energy and make it available at any time. He says, "Various companies, including GE [General Electric], are investing heavily in ways to come up with cost-effective systems to store the wind and solar energy that their wind turbines and solar technologies capture in combination with other means of generating electricity and power." But thus far, says Marcial, "Even these giants are finding difficulty in grappling with the problem."[49] Until a storage solution is found, wind power's inconsistency will prevent it from becoming a significant replacement for fossil fuel.

Electricity is essential to modern life, and researchers around the world are continually searching for cleaner and less expensive ways to produce it. Of all the current alternatives to fossil fuels, wind power is one of the best. However, the fact that wind is better than some of the alternatives does not mean it is a good replacement for fossil fuels. Says power expert George C. Loehr, "Unfortunately, whatever wind turbines can do to reduce carbon emissions, wind capacity is not the silver bullet that will solve all our future electric supply needs."[50] Wind power is too unpredictable and intermittent to viably replace significant amounts of fossil fuel.

Chapter Four

Should Government Play a Role in Developing Wind Power?

The Government Should Help Develop Wind Power

Wind power needs government support to become established and competitive in the electricity market. Government support also ensures the market stability and consistency that is necessary to attract researchers and developers to the field. By helping the wind industry, the government is not giving it special treatment; it is simply giving it the same treatment fossil fuels get. In addition, wind power is deserving of government support because it is for the greater good of society, with benefits such as reduced pollution, economic stimulation, and energy security.

The Debate

The Government Should Not Be Involved in Developing Wind Power

Government involvement in the wind power industry is both unnecessary and harmful. History shows that government involvement in any energy technology frequently has negative effects. Wind technology will develop more effectively in the free market, where market forces will decide whether it is a viable source of electricity production. Government support actually reduces innovation by allowing wind power companies to profit regardless of what they do. The harms of government support have been revealed in Europe, where a number of governments have heavily supported the development of wind power with primarily negative results.

The Government Should Help Develop Wind Power

"Government support is critical in helping businesses get new energy ideas off the ground."

—President Barack Obama, State of the Union Address, 2012. www.whitehouse.gov.

Wind power has the potential to generate a significant percentage of electricity in the United States and the world, but government support is needed to help it reach that point. Fossil fuels have been providing most of the world's electricity for many years and have had time to become firmly established in the electricity market. Wind technology will not become competitive with this existing technology without government help. The Union of Concerned Scientists believes that wind power needs a broad range of government support to become established. It says, "State and federal policies such as tax credits and other financial incentives, increased funding for research and development, and improved processes for transmission planning, siting, and approval are also needed to ensure a successful future for wind power."[51] With government support, wind power has already made substantial progress in becoming an important source of power but it needs continued support to become fully competitive with other existing sources of electricity.

A History of Government Help for Energy

Wind power is not unique among energy sources in needing government help. In the past, government subsidies have helped develop other new sources of power, including fossil fuels. Linda Taylor of Fresh Energy, a nonprofit organization that promotes the use of renewable energy, argues,

"No electricity generation technology or resource has been developed in the United States into marketable electricity without significant subsidies."[52]

In a 2011 report Nancy Pfund and Ben Healey analyzed the history of government support for various energy sources in their first fifteen years of development. They found that oil and gas received approximately half a percent of the federal budget, whereas renewable energy received only about a tenth of a percent. The researchers concluded, "Federal incentives for early fossil fuel production and the nascent nuclear industry were much more robust than the support provided to renewables today."[53]

Security Is Essential for Development

Not only does the wind power industry need government support now, it needs a guarantee that support will continue in the near future. This will give the industry the predictability, consistency, and market stability it needs to develop. By providing tax credits and other financial incentives, the government encourages investment in wind power. This investment will allow the industry to develop and eventually become competitive with other sources of power. However, if there is no guarantee that government support will continue in the future, researchers and investors are less likely to become involved in the industry. So far support for wind power in the United States has been temporary and inconsistent. According to a 2011 report by CLF Ventures and the Massachusetts Clean Energy Center, "The economic incentives, such as tax credits and grants, are periodically renewed by the federal government for specific periods of time. In the past these incentives have expired and thus have been unavailable for periods of time."[54]

One example of the harms of inconsistent government support is the way the US government has overseen the federal Production Tax Credit

> "No electricity generation technology or resource has been developed in the United States into marketable electricity without significant subsidies."[52]
>
> —Linda Taylor, the Clean Energy & Energy Efficiency director of the organization Fresh Energy.

Government Financial Support Is Needed

The federal Production Tax Credit (PTC) gives wind farms tax credit for the generation of electricity during their first 10 years of operation. While various economic factors affect industry growth, history shows that this form of government support is important to wind industry development. In those years where the tax credit was available there were increases in annual wind capacity, or the amount of wind power the United States is capable of producing. However, in 1999, 2001, and 2003 when the credit was allowed to expire, wind power installation decreased significantly. (Decreases in other years resulted from other factors.) Experts project that allowing the credit to expire again may lead to another decrease.

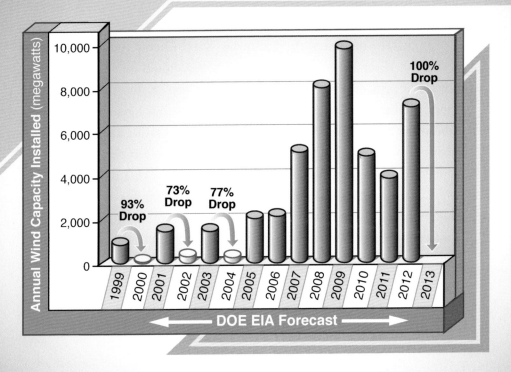

Source: American Wind Energy Association, "Federal Production Tax Credit for Wind Energy," 2011. www.awea.org.

(PTC). This is one of the main ways the federal government has supported wind power. It gives wind farms tax credit for every kilowatt hour of electricity they generate for the first 10 years of operation. However, the PTC has expired numerous times in the past and was set to expire again at the end of 2012. Journalist Stephen Lacey explains how the uncertainty over whether this credit will be available in the future hinders the development of wind power. He says, "Because it can take years to plan large wind farms, many projects are delayed or abandoned if a developer is unsure about completing the facility in time to qualify for tax credits. Lacey adds, "If the PTC expires, the wind industry would see a massive decline in installations, effectively choking one of the fastest-growing energy sectors in the country."[55] According to the American Wind Energy Association, in the past when the PTC was allowed to expire, the impact on the wind industry was significant. The association reports that PTC expiration caused wind power installations to be reduced between 73 and 93 percent and resulted in significant job losses.

In a 2010 interview, the European Wind Energy Association CEO Christian Kjaer argues that the United States should follow Europe's example of providing the wind industry with predictable, long-term support. He believes that such support in Europe has helped wind power and other renewables develop more quickly. He says, "The U.S. framework for investing in renewables is very unstable—I mean, it cannot be predicted more than one or two years ahead. And that also means that the United States is not reaping the job creation benefits of wind energy, because a lot of components, a lot of manufacturing is imported because no one's going to invest in a factory in the United States if they don't know how the market looks beyond the next two years." In Kjaer's opinion, "What has given rise to those [EU] markets [for wind power and other renewables] is that you have stable frameworks and they have been long-term. The problem in the U.S. is that the framework expires every year or every second year."[56]

Equality with Fossil Fuels

Government support for private industry is unpopular these days, but it is important to note that the fossil fuel industry has benefited hugely from such support. Fossil fuels currently receive significant government subsidies and other help, and they have for many years. According to the US Energy Information Administration, in 2010 coal received $1.4 billion in government subsidies and other support, and natural gas and petroleum liquids received

$2.8 billion. Although wind power is also receiving significant government support at present—$4.9 billion in 2010—it has received minimal support in the past. According to CLF Ventures and the Massachusetts Clean Energy Center, between 2002 and 2008 fossil fuels received approximately $72 billion in subsidies, whereas wind power and other renewables *combined* only received about $12 billion.

In addition to a long history of government support, fossil fuels are favored in the way they receive that support. A significant part of the fossil fuel subsidies comes from permanent tax breaks written into the US tax code, whereas support for wind and other renewables is temporary and must be renewed periodically by the government. Jeff Deyette, an energy analyst for the Union of Concerned Scientists, argues that it is difficult for wind to compete in the electricity market without equal treatment. "We've had decades of incentives that have skewed the market toward natural gas and other fossil fuels," he says.[57] To grow, wind energy needs a level playing field.

> "We've had decades of incentives that have skewed the market toward natural gas and other fossil fuels."[57]
>
> —Jeff Deyetter, an energy analyst for the Union of Concerned Scientists.

Every year, the fossil fuel industry makes millions of dollars in political contributions to the government, and as a result the industry continues to receive more favorable treatment than the wind power industry. According to a 2011 *Time* magazine article, the fossil fuel industry spends $200 million a year in lobbying and political contributions that help influence favorable government treatment. By comparison, says *Time*, in 2010 clean energy lobbyists spent about $30.7 million. Because it cannot afford to spend as much money on lobbying as the fossil fuel industry, the wind power industry continues to receive less favorable government treatment than fossil fuels.

For the Greater Good

Finally, the government should support wind power because the wind power industry will benefit society in more ways than just providing electricity. Wind power also reduces pollution, creates jobs, and reduces dependence on foreign energy. Government support will ensure that wind power succeeds and these important social benefits are realized. In the United States and many other countries, the government often supports ventures that are judged to be for the greater good of society and that might not otherwise come into existence. For example, it helps build mass transit systems that reduce the number of vehicles on the road, thus reducing fuel consumption and pollution. Because wind power is an important public good, government should ignore arguments about affordability and comparisons to fossil fuels, and simply focus on doing what it needs to do to ensure the success of this technology. As Taylor explains, "What is critical for new electricity generation technologies—and those that are not so new but that are newly moving towards marketability—is to determine the likelihood that they carry significant benefits to the system and, if so, what it will take in additional economic incentives and subsidies to make them viable."[58] Wind power has significant public benefits and government should give it the subsidies it needs to become viable.

In an examination of wind power, journalist Bryan Walsh looks at the example of Denmark, which is a world leader in wind power. He points out that this did not happen simply through the actions of the free market. Walsh argues, "Denmark has been a success in wind power because it wanted to be."[59] It did this through government action in the form of subsidies and loans for wind power and mandates on utilities to purchase wind power. If other countries want to be successful in developing wind power in the coming years, they will need to follow Denmark's example and use the power of the government to help this industry develop.

The Government Should Not Be Involved in Developing Wind Power

"It is time that wind pulls its own weight instead of relying on taxpayers' dollars and additional subsidies and mandates."

—Daniel Simmons, director of state and regulatory affairs at the Institute for Energy Research.

Quoted in David Hosansky, "Wind Power: Is Wind Energy Good for the Environment?," *CQ Researcher*, April 1, 2011. http://library.cqpress.com.

Government involvement in wind power harms its development by distorting free market forces. The free market is better than government policy to guide wind power development because in the free market, only those technologies that are economically viable will succeed. Margo Thorning, chief economist of the American Council for Capital Formation, believes that the government should not fund risky wind power and other renewable energy projects. She says, "The government should limit its involvement and funding to basic research on alternative energy sources and should not be funding risky 'start-ups.' If a renewable technology [such as wind power] makes economic sense, the private sector will adopt it and it will succeed without mandates and subsidies."[60] Wind power should be left to the power of the free market—without government support—to determine whether it really is a worthwhile technology for society. As Republican representative Mike Pompeo of Kansas argues, "Wind, solar, biomass and other sources of energy have all shown great promise, but it is high time for energy sources to demonstrate their value on the open market—without government interference."[61]

Government Mandates Fail

In contrast, when the government creates mandates and subsidies for certain technologies—in this case, wind power—it may be encouraging the development of something that does not actually make practical or economic sense. Frank M. Stewart, president and chief executive officer of the American Association of Blacks in Energy, points out that many of the important technological innovations of the past occurred as a result of free market forces, not government intervention. He says, "From Thomas Edison inventing electric-power generation in 1880 to Elijah McCoy a decade earlier creating the automatic lubricator necessary to grease the steam engines of trains . . . America's history is filled with pioneers and entrepreneurs who have pushed our economy forward through private innovation."[62]

> "Wind and solar technology in their current form cannot compete in the marketplace without heavy government intervention, and thanks to that intervention, wind and solar purveyors don't *have* to innovate to make a profit."[68]
>
> —Ed Morrissey, senior editor for the news and commentary site Hot Air.

Journalist Steven Mufson analyzes the history of government involvement in the energy industry and finds that it has not been beneficial. According to Mufson, government investment in this field has led to "a graveyard of costly and failed projects."[63] For example, says Mufson, between 2004 and 2008 the government invested $1.2 billion in developing hydrogen vehicles, only to have the project fail. Likewise, he says, hundreds of millions of dollars have been invested in developing clean coal technology—which involves capturing and burying the emissions generated from burning coal—with little success. Overall, says Mufson, "Not a single one of these much-ballyhooed initiatives is producing or saving a drop or a watt or a whiff of energy, but they have managed to burn through . . . [large amounts of] taxpayer money."[64]

An Inferior Technology

The argument that government funding is needed because wind power is not yet a mature technology and cannot compete with other energy sources until it develops fully, does not hold up. Society has utilized wind power for many years and the real reason it cannot compete with industries like coal is that it is inferior. Daniel Simmons, director of state and regulatory affairs for the Institute of Energy Research argues, "Wind may not seem like a 'mature' technology because it is unreliable. People switched away from wind to other sources of energy such as coal, hydro-electric, natural gas and petroleum because the wind doesn't always blow, and these other sources could be counted on."[65]

Even with many years to develop and significant government support, wind power has not proven to be a viable technology. Instead of continuing to support it, the government should invest in other energy technologies that have more promise. Nathan Myhrvold, former chief strategist and chief technology officer at Microsoft Corporation argues, "Rather than accelerate deployment of technologies that we know are inefficient, wouldn't it be better to invest in the research and development that are needed to come up with renewable technologies that are cheap as well as clean?"[66]

Financial Incentives Prevent Innovation

Government involvement in wind power is actually holding back the development of wind technology. In the United States, government support for wind power is mainly through tax credits for wind power producers, grants for research and development, and mandates requiring that a certain percentage of electricity comes from renewable energy sources such as wind. Many companies spin off into renewable energy ventures such as wind power simply to take advantage of these incentives rather than to create an effective, cost-competitive product. Power generation technology expert Ed Hiserodt says, "These companies [investing in wind power] are not so much interested in creating power, but in siphoning government subsidies and taking advantage of 'renewable' energy tax breaks."[67]

In most cases the power produced by the wind is still too expensive to result in a profit for investors, so wind power companies are not profitable unless they receive government financial support. However, as Ed Morrissey, senior editor for the news and commentary site Hot Air, argues, because government subsidies make the wind power industry profitable, there is no incentive for wind power to develop. He says, "Wind and solar technology in their current form cannot compete in the marketplace without heavy

Federal Subsidies Cannot Be Justified

Electricity generated from the wind is generally more expensive than that generated from fossil fuels. As a result, many people argue that subsidies for wind power are a poor investment of government funds. This graph shows that compared to other sources of electricity, wind power produces a far smaller amount of electricity relative to the subsidies it receives. The graph compares the federal subsidies for various types of electricity generation relative to the amount of electricity actually produced by each. It reveals that wind power receives $56.29 in federal subsidies for every megawatt hour of electricity it produces, while coal produces a megawatt hour with a subsidy of only $0.64. Megawatt-hours are a common way to measure the production of electrical power.

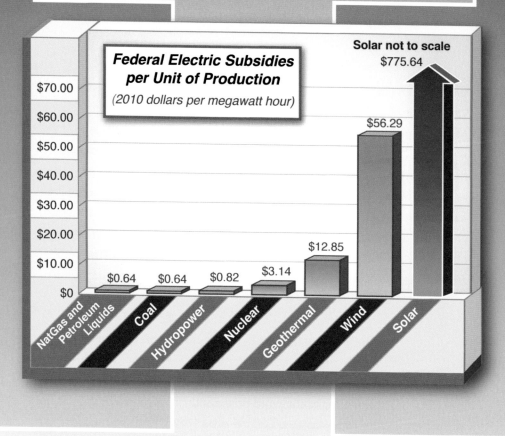

Federal Electric Subsidies per Unit of Production

(2010 dollars per megawatt hour)

Solar not to scale $775.64

$56.29 — Wind

$12.85 — Geothermal

$3.14 — Nuclear

$0.82 — Hydropower

$0.64 — Coal

$0.64 — NatGas and Petroleum Liquids

Source: Institute for Energy Research, "EIA Releases New Subsidy Report: Subsidies for Renewables Increase 186 Percent," August 3, 2011. www.instituteforenergyresearch.org.

government intervention, and thanks to that intervention, wind and solar purveyors don't *have* to innovate to make a profit. Morrissey adds, [Government is] preventing—or at least disincentivizing—the normal innovative processes that could unlock a real revolution in green technology."[68]

Government support for wind power also prevents innovation by diverting precious funds from more promising sources of energy. Renewable energy technologies like wind power require large government investments to make them viable, but as energy expert Jeffrey Leonard explains, they also require continuing investment to help them survive in the market, and this means the government has no money to invest in other sources of energy. He warns, "Once built, these boondoggles create political interests demanding still more subsidies while limiting our nation's flexibility to shift to cleaner and more cost-efficient energy sources."[69]

Failure of Government Involvement in Europe

In countries where the government has been heavily involved in the development of wind power, there are reports of harmful economic effects. Government subsidies and other support have helped wind power grow in many European countries, including Netherlands, Germany, Spain, and Denmark. These subsidies, in addition to policies that mandate the use of wind energy and other renewable sources, have helped the wind power industry grow and created new jobs as a result. However, there have been negative effects on other parts of the economy. For example, using more wind power has increased the overall price of electricity for consumers because it costs more to produce electricity from the wind than from fossil fuels. And even though jobs have been added in the wind power industry, the money used to create those jobs could be invested in other sectors of the economy to create even more jobs. In addition, some wind power jobs have been created by taking jobs from other parts of the economy. Thus, wind power–related jobs are actually reducing the total number of jobs available.

Kenneth P. Green, a scholar at the American Enterprise Institute, argues that the example of Europe reveals that government intervention in wind power and other renewables is harmful overall. He maintains that as a result of government involvement, energy prices have increased significantly, and the economy has been harmed through lost jobs. Says Green, "Green programs in Spain destroyed 2.2 jobs for every green job created, while the capital needed for one green job in Italy could create almost five jobs in the general economy." In addition, he says, "Wind and solar power have raised household energy prices by 7.5 percent in Germany, and Denmark has the highest electricity prices in the European Union."[70]

Germany is an example of the harms of government involvement in the wind power industry. The German government has aggressively supported wind power and other types of renewable energy and is a world leader in the production of wind energy. However, this support has not had the desired result of allowing the wind industry to become competitive and profitable. Instead, the result has been high household energy prices and a wind power industry that needs continual government support. In an analysis of Germany's policies, researchers with a German energy research institute conclude, "The government's support mechanisms have in many respects subverted these incentives, resulting in massive expenditures that show little long-term promise for stimulating the economy, protecting the environment, or increasing energy security."[71]

> "Wind and solar power have raised household energy prices by 7.5 percent in Germany, and Denmark has the highest electricity prices in the European Union."[70]
>
> —Kenneth P. Green, a scholar at the American Enterprise Institute.

Government Support Makes No Sense

The government must be selective in the way it uses its resources. It does not have the ability to pay for everything. Supporting wind power should not be one of government's responsibilities, because it will be both ineffective and harmful to the future development of the wind power industry.

Source Notes

Overview: Visions of the Future: Wind Power

1. Quoted in Andrés Cala, "Tiny Spanish Island Has a Huge Stake in the Future," *New York Times*, January 19, 2011. www.nytimes.com.
2. Kate Galbraith, "Wind Power Gains as Gear Improves," *New York Times*, August 7, 2011.
3. US Department of the Interior Bureau of Land Management, "Frequently Asked Questions," Wind Energy Development Programmatic Environmental Impact Statement. http://windeis.anl.gov.
4. Peter Musgrove, *Wind Power*. Cambridge, UK: Cambridge University Press, 2010, p. 11.

Chapter One: Is Wind Power Affordable?

5. Union of Concerned Scientists, "Tapping into Wind Power," April 2011. www.ucsusa.org.
6. Centre for Sustainable Energy, "Common Concerns About Wind Power," May 2011. www.cse.org.
7. Canadian Wind Energy Association, "Wind Facts: Pricing," 2011. www.canwea.ca.
8. Canadian Wind Energy Association, "Wind Facts: Pricing."
9. Chris Nelder, "The End of Fossil Fuel," *Forbes*, July 24, 2009.
10. US Department of Energy, "Wind Compared to the Cost of Other Electricity Generating Options," New England Wind Forum, March 3, 2011. www.windpoweringamerica.gov.
11. Philip Totaro, "Q&A: Has Wind Energy Reached Overcapacity?," Composites Manufacturing, August 30, 2011. www.compositesmanu facturingblog.com.
12. Natural Gas Council, "Natural Gas Council Statement on State of the Union Speech," Washington, DC, January 25, 2012. www.aga.org.
13. K. de Groot and C. le Pair, "The Hidden Fuel Costs of Wind Generated Electricity," National Wind Watch, December 18, 2009. www .wind-watch.org.
14. Benjamin Zycher, "Wind and Solar Power, Part II: How Persuasive Are the Rationales?," *American Enterprise Institute*, January 17, 2012. www.aei.org.

15. Glenn Schleede, "Wind Farms DO NOT Provide Large Economic and Job Benefits (Quite the Opposite)," *MasterResource: A Free-Market Energy Blog*, January 5, 2011. www.masterresource.org.

16. De Groot and Le Pair, "The Hidden Fuel Costs of Wind Generated Electricity."

Chapter Two: How Does Wind Power Impact the Environment?

17. US Department of Energy, "20% Wind Energy by 2030: Increasing Wind Energy's Contribution to U.S. Electricity Supply," December 2008. www1.eere.energy.gov.

18. Dora Anne Mills, "In Opposition to Rule Changes to Maine DEP. Chapter 375 Regulations on Wind Turbine Noise," testimony, Maine Board of Environmental Protection public hearing, July 7, 2011. www.maine.gov/dep/ftp/bep/ch375citizen_petition/pre-hearing/AR-83%20-%20Dora%20Mills%20testimony.pdf.

19. Union of Concerned Scientists, "Tapping into Wind Power."

20. Quoted in American Wind Energy Association, "World Water Day: Wind Power Saves Water," March 22, 2011. www.awea.org.

21. Centre for Sustainable Energy, "Common Concerns About Wind Power."

22. W. David Colby et al., "Wind Turbine Sound and Health Effects: An Expert Panel Review," American Wind Energy Association and Canadian Wind Energy Association, December 2009. www.canwea.ca.

23. Quoted in Kristin Choo, "The War of the Winds: Wind Farms Are a Growing Source of Clean Energy. But Some of the Neighbors Are Beginning to Complain," *ABA Journal*, February 2010. www.abajournal.com.

24. Robert Bryce, *Power Hungry: The Myths of "Green" Energy and the Real Fuels of the Future*. New York: PublicAffairs, 2010, p. 84.

25. Bryce, *Power Hungry*, p. 95.

26. *National Review*, "Britain's Experience with Wind Power Illustrates How Engineering Truths Always Triumph over Green Dreams," August 1, 2011. www.nationalreview.com.

27. Bentek Energy, "How Less Became More: Wind Power and Unintended Consequences in the Colorado Energy Market," April 16, 2010. www.bentekenergy.com.

28. Quoted in Katharine Lochnan Balaclava, "Scenic Beauty Worth Preserving," *Owen Sound (ON) Sun Times*, January 4, 2011. www.owensoundsuntimes.com.

29. Chad Pepin, "NIMBY and PROUD!," *Wind Turbine Syndrome Resources*, September 1, 2010. www.windturbinesyndrome.com.

30. Quoted in Robert Johns, "Wind Power Could Kill Millions of Birds Per Year by 2030," *American Bird Conservancy*, February 2, 2011. www.abcbirds.org.

31. Erich Schwartzel, "Pennsylvania Wind Turbines Deadly to Bats, Costly to Farmers," *Pittsburgh Post-Gazette*, July 17, 2011. www.post -gazette.com.

32. Quoted in Choo, "The War of the Winds."

33. Jane Davis, "General Statement by Jane & Julian Davis," *European Platform Against Windfarms*, April 16, 2009. www.epaw.org.

Chapter Three: Can Wind Power Ever Replace Fossil Fuels?

34. David J.C. MacKay, "Illuminating the Future of Energy," *New York Times*, August 28, 2009. www.nytimes.com.

35. Quoted in Wei Tong, *Wind Power Generation and Wind Turbine Design*. Southampton, UK: WIT, 2010, p. 363.

36. Cape Wind, "Project at a Glance," www.capewind.org.

37. Quoted in O. Edenhofer et al., eds., *IPCC Special Report on Renewable Energy Sources and Climate Change Mitigation*. Cambridge, UK: Cambridge University Press, 2011, pp. 82–83.

38. American Wind Energy Association, "How Wind Energy Is Reliably Integrated on the Grid," May 2011. www.awea.org.

39. Musgrove, *Wind Power*, pp. 220–21.

40. US Department of Energy, "20% Wind Energy by 2030."

41. Bryce, *Power Hungry*, pp. 4–5.

42. Bjorn Lomborg, "Done with the Wind," *Newsweek International*, March 21, 2011. www.newsweek.com.

43. Richard Stieglitz with Rick Docksai, "Why the World May Turn to Nuclear Power," *Futurist*, November/December 2009. www.wfs.org.

44. Stop Ill Wind, "Today's Challenge for Responsible Wind Citizenship." www.stopillwind.org.

45. Ed Hiserodt, "Wind Power: An Ill Wind Blowing," *New American*, October 27, 2010. www.thenewamerican.

46. Jon Boone, "Why Wind Won't Work," Save Western Ohio, April 26, 2008. http://savewesternoh.org.

47. Ted Sickinger, "Too Much of a Good Thing: Growth in Wind Power Makes Life Difficult for Grid Managers," *OregonLive.com*, July 20, 2011. www.oregonlive.com.

48. Bryce, *Power Hungry*, pp. 105–106.

49. Gene Marcial, "Power Opportunity for GE, et al.: How to Store Wind/Solar Renewable Energy," *Forbes*, May 2, 2011. www.forbes.com.

50. George C. Loehr, "The Trouble with Wind," *Transmission & Distribution World*, May 1, 2011. http://tdworld.com.

Chapter Four: Should Government Play a Role in Developing Wind Power?

51. Union of Concerned Scientists, "Tapping into Wind Power."

52. Linda Taylor, "The Tiresome Grousing over Energy Subsidies," Fresh Energy, October 6, 2011. http://fresh-energy.org.

53. Nancy Pfund and Ben Healey, "What Would Jefferson Do? The Historical Role of Federal Subsidies in Shaping America's Energy Future," DBL Investors, September 2011, p. 6.

54. CLF Ventures and the Massachusetts Clean Energy Center, "Land-Based Wind Energy: A Guide to Understanding the Issues and Making Informed Decisions," June 2011. http://masscec.com.

55. Stephen Lacey, "Harry Reid: I'm 'Not Confident' Congress Can Extend the Production Tax Credit for Wind," *ThinkProgress*, September 6, 2011. http://thinkprogress.org.

56. Quoted in Fen Montaigne (interview), "Wind Power's Growth Is Blowing Europe Toward Green Goals," *Guardian*, September 9, 2010. www.guardian.co.uk.

57. Quoted in David Hosansky, "Wind Power: Is Wind Energy Good for the Environment?," *CQ Researcher*, April 1, 2011. http://library.cqpress.com.

58. Taylor, "The Tiresome Grousing over Energy Subsidies."

59. Bryan Walsh, "The Gusty Superpower," *Time*, March 16, 2009. www.time.com.

60. Margo Thorning, "Stop DOE's Double Down on Risky Energy Ventures," *National Journal Expert Blogs: Energy & Environment*, September 29, 2011. http://energy.nationaljournal.com.

61. Quoted in Rob Hotakainen, "Congress Divided over Continuing Subsidization of Wind Power," *McClatchy Newspapers*, November 14, 2011. www.mcclatchydc.com.

62. Frank M. Stewart, "Solyndra vs. Shipyards in Energy, Jobs," *National Journal Expert Blogs: Energy & Environment*, September 30, 2011. http://energy.nationaljournal.com.

63. Steven Mufson, "Before Solyndra, a Long History of Failed Government Energy Projects," *Washington Post*, November 11, 2011. www.washingtonpost.com.

64. Mufson, "Before Solyndra, a Long History of Failed Government Energy Projects."

65. Quoted in Hosansky, "Wind Power."

66. Nathan Myhrvold, "Energy Subsidies Stymie Wind, Solar Innovation: Nathan Myhrvold," *Bloomberg Business Week*, November 29, 2011. www.businessweek.com.

67. Ed Hiserodt, "Wind Power II: The Wind-Farm Eruption," *New American*, October 28, 2010. http://thenewamerican.com.

68. Ed Morrissey, "Reason: Should Government Be Tilting at Wind Turbines?," Hot Air, March 12, 2011. http://hotair.com.

69. Jeffrey Leonard, "Get the Energy Sector off the Dole," *Washington Monthly*, January/February 2011. www.washingtonmonthly.com.

70. Kenneth P. Green, "The Myth of Green Energy Jobs: The European Experience," *American Enterprise Institute*, February 15, 2011. www.aei.org.

71. Manuel Frondel, Nolan Ritter, and Colin Vance, "Economic Impacts from the Promotion of Renewable Energies: The German Experience," Rheinisch-Westfälisches Institut für Wirtschaftsforschung, October 2009. www.instituteforenergyresearch.org.

Wind Power Facts

Production of Wind Power

- According to the Global Wind Energy Council, in 2010 half of all new global wind installations were in China.
- According to the American Wind Energy Association, in 2010 Texas had the largest installed wind power capacity—the amount of wind power it was capable of producing—in the country, followed by Iowa, then California. Texas's capacity was more than double that of any other state.
- The US Energy Information Administration reports that in 2010 approximately 45 percent of US electricity came from coal, 24 percent from natural gas, and 2 percent from wind power.
- According to the European Wind Energy Association, the largest onshore wind turbine in Europe is 440 feet (134m) tall, and its blades sweep an area 416 feet (127m) in diameter.
- In 2011 article the *New York Times* reports that one wind turbine has approximately 7,000 components.

Industry Growth

- According to a report prepared by the Lawrence Berkeley National Laboratory for the DOE, in 2010 wind power accounted for 25 percent of additions to the electricity generating capacity of the United States, down from 42 percent in 2009.
- The Center for Climate and Energy Solutions estimates that with aggressive development, wind power could account for as much as 13 percent of global electricity production by 2035.
- The European Wind Energy Association predicts that by 2030, Europe will meet 28 percent of its electricity demand with wind power.
- According to the American Wind Energy Association, in 2010, 38 US states had utility-scale wind installations—installations that are large enough to provide power for public utilities.

- The Danish Wind Industry Association reports that in 2010, China installed the majority of the world's new wind power capacity.

Economics of Wind Power

- According to the Center for Climate and Energy Solutions, over the past few decades, there has been a significant reduction in the cost of wind power, with wind turbines producing 30 times as much power as they did 30 years ago, at a much lower cost.
- The DOE estimates that meeting the goal of supplying 20 percent of the country's electricity through wind power by 2030 would create 500,000 jobs directly in the wind industry, 100,000 in associated industries such as electrical manufacturing and steel, and 200,000 through economic expansion as a result of local spending.
- The market research company IBISWorld predicts that between 2000 and 2016, wind power will be the second-fastest-growing industry in the United States.
- The Center for Climate and Energy Solutions reports that offshore wind turbines are about 50 percent more expensive than onshore ones but produce about 50 percent more electricity.
- According to the American Wind Energy Association, wind power helps the US economy, with over 400 American manufacturing plants building components for wind turbines.
- In a 2011 report for the Manhattan Institute, energy expert Robert Bryce estimates that for the United States to get 20 percent of its electricity from the wind by 2030 would cost more than $850 billion.

Wind Power and the Environment

- The Environmental Protection Agency estimates that electricity generation, which is primarily from fossil fuels, accounts for about a third of US global warming emissions.
- According to a 2010 report by the DOE, the wind turbines currently operating in the United States reduce carbon dioxide emissions by 62 million tons (56.2 million metric tons) each year (the same as taking

10.5 million cars off the road) and save 20 billion gallons (75.7 billion L) of water that would otherwise be used in conventional power plants.

- The American Bird Conservancy estimates that by 2030, wind farms in the United States will impact almost 20,000 square miles (51,800 sq. km) of land that is a habitat for birds.
- Many people object to wind power because of its impact on animal and human health and on the environment. According to Robert Bryce of the Manhattan Institute, the United States has approximately 170 anti–wind power groups, the United Kingdom about 250, and Europe more than 400.
- The National Wind Coordinating Collaborative reports that reducing turbine operation during low wind periods could reduce bat fatalities by 50 to 87 percent, with only a modest reduction in power production.

Related Organizations and Websites

American Wind Energy Association (AWEA)
1501 M St. NW, Suite 1000
Washington, DC 20005
phone: (202) 383-2500 • fax: (202) 383-2505
e-mail: windmail@awea.org • website: www.awea.org

The AWEA is a trade organization with over 2,500 members and advocates, including project developers, equipment suppliers, and researchers. Its website provides current statistics on wind power and information about wind power policies and projects.

Canadian Wind Energy Association (CanWEA)
1600 Carling Ave., Suite 710
Ottawa, ON K1Z 1G3 Canada
phone: (613) 234-8716 • fax: (613) 234-5642
website: www.canwea.ca

CanWEA is a nonprofit trade association that promotes the appropriate development and application of all aspects of wind energy in Canada. It was established in 1984, and its membership includes individuals involved in all facets of wind energy development. Its website has general information about wind energy and statistics about wind energy development in Canada.

European Platform Against Windfarms (EPAW)
3 Rue des Eaux
75016 Paris, France
phone: 33 680 99 38 08
e-mail: contact@epaw.org • website: http://epaw.org

The EPAW has 502 member organizations from 22 countries and works to help its members oppose wind farms or question their effectiveness and fight against the damaging effects of wind farms. Its website has links to many publications about wind power and numerous testimonies about the negative effects of wind farms.

European Wind Energy Association (EWEA)
Rue d'Arlon 80
B-1040 Brussels, Belgium
phone: 32 2 213 1811 • fax: 32 2 213 1890
e-mail: ewea@ewea.org • website: www.ewea.org

The EWEA is a large organization, with members in almost 60 countries, that promotes the use of wind power in Europe and around the world. It analyzes and formulates policy about important wind power issues and cooperates with research projects.

National Renewable Energy Laboratory (NREL)
1617 Cole Blvd.
Golden, CO 80401
phone: (303) 275-3000
website: www.nrel.gov

The NREL is the DOE's laboratory for renewable energy research and development. Its website has maps, graphs, charts, and reports about renewable energy, including wind power.

National Wind Coordinating Collaborative (NWCC)
1255 Twenty-Third St. NW, Suite 875
Washington, DC 20037
website: www.nationalwind.org

The NWCC is made up of representatives from both the public and private sector and has the goal of developing wind power. Its website has current news about wind power and numerous statistics, reports, and fact sheets.

Society for Wind Vigilance
phone: (647) 588-8647
e-mail: communications@windvigilance.com
website: www.windvigilance.com

The Society for Wind Vigilance is an international federation that aims to protect the health and safety of communities as wind turbine construction proliferates around the world. Its website contains information about the potential harms of wind turbines and advocates additional research in relation to placement of wind turbines due to these potential harms.

World Wind Energy Association
Charles-de-Gaulle-Str. 5
53113 Bonn, Germany
phone: 49 228 369 40 80 • fax: 49 228 369 40 84
website: www.wwindea.org

The World Wind Energy Association is a nonprofit organization that works to promote the deployment of wind energy around the world. It has members in 95 countries. The organization believes that wind energy is an important part of replacing fossil fuels. Its website has statistics and publications about wind energy.

For Further Research

Books

Robert Bryce, *Power Hungry: The Myths of "Green" Energy and the Real Fuels of the Future*. New York: PublicAffairs, 2010.

Dan Chiras, *Wind Power Basics*. Gabriola Island, BC: New Society, 2010.

John Etherington, *The Wind Farm Scam*. London: Stacey, 2009.

Larry Flowers and Sandra Reategui, *Wind Power Answer in Times of Water Scarcity*. Golden, CO: National Renewable Energy Laboratory, 2010.

Roger E. Meiners, et al., eds., *The False Promise of Green Energy*. Washington, DC: Cato Institute, 2011.

Peter Musgrove, *Wind Power*. Cambridge, UK: Cambridge University Press, 2010.

Periodicals

Kristin Choo, "The War of the Winds: Wind Farms Are a Growing Source of Clean Energy. But Some of the Neighbors Are Beginning to Complain," *ABA Journal*, February 2010.

Erik Heinrich, "Breezing In," *Time*, June 20, 2011.

David Hosansky, "Wind Power: Is Wind Energy Good for the Environment?," *CQ Researcher*, April 1, 2011.

Jeffrey Leonard, "Get the Energy Sector off the Dole," *Washington Monthly*, January/February 2011.

Andrew Walden, "Wind Energy's Ghosts," *American Thinker*, February 15, 2010.

Internet Sources

American Wind Energy Association and Solar Energy Industries Association, "Green Power Superhighways: Building a Path to America's Clean Energy Future," 2009. www.awea.org/documents/issues/upload/Green PowerSuperhighways.pdf.

CLF Ventures and the Massachusetts Clean Energy Center, "Land-Based Wind Energy: A Guide to Understanding the Issues and Making Informed Decisions," June 2011. http://masscec.com/masscec/file/Wind _Guide(1).pdf.

W. David Colby et al., "Wind Turbine Sound and Health Effects: An Expert Panel Review," American Wind Energy Association and Canadian Wind Energy Association, December 2009. www.canwea.ca/pdf /talkwind/Wind_Turbine_Sound_and_Health_Effects-Executive_Sum mary.pdf.

Society for Wind Vigilance, "An Analysis of the American/Canadian Wind Energy Association Sponsored 'Wind Turbine Sound and Health Effects: An Expert Panel Review, December 2009,'" January 2010. www .windvigilance.com.

Union of Concerned Scientists, "Tapping into Wind Power," April 2011. www.ucsusa.org/assets/documents/clean_energy/tappingintothewind .pdf.

US Energy Information Administration, "Annual Energy Outlook 2011: With Projections to 2035," April 2011. www.eia.gov/forecasts/aeo /pdf/0383(2011).pdf.

Index

Note: Boldface page numbers indicate illustrations.